This study of Panhandle Eastern's history explores the relationship between regulatory policy and the modern corporation in the twentieth century from a unique perspective, for it extends over three eras in the growth of the pipeline industry and the U.S. political economy. The first era, in which the interstate pipeline industry began, was characterized by cartels, stock manipulation, territorial agreements, and espionage with minimal regulation or antitrust enforcement. Then, New Deal regulatory reforms broke up the utility combines and subjected interstate pipeline firms to single industry regulation. Pipelines bought and sold gas and enjoyed a long period of expansion during this second era in spite of the increasingly complex regulations imposed on them. Finally, the third era was characterized by regulatory failure, energy crises, regulatory change, innovation, and industry reorganization as regulators removed the traditional merchant function, that of buying and selling gas, from pipelines and transformed them into open access contract carriers.

Gas pipelines and the emergence of America's regulatory state

STUDIES IN ECONOMIC HISTORY AND POLICY
THE UNITED STATES IN THE TWENTIETH CENTURY

Edited by
Louis Galambos and Robert Gallman

Gas pipelines and the emergence of America's regulatory state

A history of Panhandle Eastern Corporation, 1928–1993

Christopher J. Castaneda
California State University
Sacramento

Clarance M. Smith

CAMBRIDGE
UNIVERSITY PRESS

Published by the Press Syndicate of the University of Cambridge
The Pitt Building, Trumpington Street, Cambridge CB2 1RP
40 West 20th Street, New York, NY 10011-4211, USA
10 Stamford Road, Oakleigh, Melbourne 3166, Australia

First published 1996

Printed in the United States of America

Library of Congress Cataloging in Publication has been applied for.

A catalog record for this book is available from the British Library

ISBN   0-521-56166-3 Hardback

Dedicated to
Joseph A. Pratt
teacher, mentor, and friend

Contents

Illustrations

Charts

Figures

Maps

Tables

Editors' preface

The regulatory state has been a major issue in American political economy for over a century. For many decades, the politics of regulation were concerned largely with determining where, by whom, and in what precise manner regulation needed to be extended to economic activities previously controlled only by market forces. The resulting political skirmishes gradually increased the area of federal and state control, largely through a piecemeal process of single-industry, rate-of-return regulation, supplemented from time to time by such cross-industry regulations as those imposed on the securities markets.

This is the context in which Christopher J. Castaneda and Clarance M. Smith place the first era of their history of Panhandle Eastern Corporation. The company was created in a laissez-faire setting before regulation was imposed by the federal government. They analyze with care the competition that ensued, the problems that existed for this fledgling enterprise, and the drives toward regulation and the application of the antitrust laws to this industry. As they show, entrepreneurs in the gas industry used all of the political, legal, and economic weapons they could bring to bear as they struggled for positions in the major urban markets.

From the New Deal through the 1960s, Panhandle built a successful business under a relatively stable regulatory regime. This was a great era of economic and technological accomplishments for the company and the industry. The political economy of abundance during the so-called American Century encouraged Panhandle and other firms in the industry to diversify, seeking opportunities outside of their core business in natural gas. Meanwhile, political and legal pressures gradually extended the regulatory regime, tightening the controls on this vital source of energy.

The result, as the authors demonstrate, was a revolution in the political economy of natural gas. The poor performance under regulation sparked a successful drive to free major elements of the industry from political control. Market forces began in these years to reshape the leading firms in natural gas. Panhandle Eastern played a major role in these significant developments and the authors show how the firm's leaders developed a series of new strategies to cope with these and the other challenges the company faced. We are extremely pleased to add this important volume to *Studies in Economic History and Policy: The United States in the Twentieth Century*.

Author's preface

In the spring of 1989, Panhandle Eastern Corporation purchased Texas Eastern Corporation for $3.2 billion making this
the highest value natural gas pipeline transaction in U.S.
history.[1] During the subsequent process of consolidating the
two firms' operations, Panhandle's James W. Hart, Jr., VP
Public Affairs, expressed to Joseph A. Pratt and Christopher
J. Castaneda the interest of the firm's CEO, Dennis Hendrix,
and management in supporting a project to document the history of Panhandle. At the time, Castaneda and Pratt were
finishing a history (begun in 1987) of the recently acquired
Texas Eastern firm and Castaneda was completing a Ph.D.
dissertation on the post–World War II expansion of the natural gas industry.[2] Panhandle's management cooperated with
the authors in their effort to finish the Texas Eastern study

---

[1] Panhandle Eastern Corporation began doing business as PanEnergy Corp on
January 2, 1996 to reflect its move into diversified energy operations.

[2] Christopher J. Castaneda and Joseph A. Pratt, *From Texas to the East: A
Strategic History of Texas Eastern Corporation* (College Station: Texas A&M
University Press, 1993). The dissertation was published subsequently as
Christopher James Castaneda *Regulated Enterprise: Natural Gas Pipelines
and Northeastern Markets, 1938–1954* (Columbus: Ohio State University
Press, 1993).

and then commenced negotiations with us for a similar history of Panhandle Eastern.

Panhandle's history spans virtually the entire scope of the modern natural gas industry, and the pipeline firm played a significant role in many of the most important legislative and regulatory decisions affecting the twentieth century gas industry, an industry which has evolved through successive periods of the absence of federal regulation, increasingly stringent regulations, and deregulation. It became increasingly clear that the history of this one natural gas pipeline firm could be used to delineate an instructive perspective on the evolution of U.S. federal regulatory policy in the twentieth century. Subsequently, we modified our research focus – with encouragement from the firm – from that of a more traditional corporate history to one which illuminated the historical political economy of an important business within a significant though largely unstudied industry.[3] To assist us in this process, Professor Louis Galambos gave the manuscript a stiff editing.

To get work started, Pratt and Castaneda, who along with Galambos are members of the Business History Group (BHG), proposed a project to research and write Panhandle's history under the same contractual arrangement they had with Texas Eastern. After brief negotiations between BHG and Panhandle Eastern, we agreed to undertake a two-phase project covering the firm's development through the twentieth century. The first phase included a one-year period for conducting oral history interviews and surveying historical corporate records. Clarance Smith, a graduate student at the University of Houston, interviewed approximately seventy retirees and employ-

---

[3] The few books which provide a historical context for the development of the natural gas industry include Alfred M. Leeston, John A. Crichton, and John C. Jacobs, *The Dynamic Natural Gas Industry: The Description of an American Industry from the Historical, Technical, Legal, Financial and Economic Standpoints* (Norman: University of Oklahoma Press, 1963); Arlon R. Tussing and Connie C. Barlow, *The Natural Gas Industry: Evolution, Structure, and Economics* (Cambridge: Ballinger Publishing Company, 1984); Elizabeth M. Sanders, *The Regulation of Natural Gas: Policy and Politics, 1938–1978* (Philadelphia: Temple University Press, 1981); Louis Stotz and Alexander Jamison, *History of the Gas Industry* (New York: Stettiner Brothers, 1938); and Malcolm W. W. Peebles, *Evolution of the Gas Industry* (New York University, 1980).

ees and surveyed existing company historical records. He was assisted on a part-time basis by company retiree J. M. "Mack" Price, Castaneda, and Pratt.

At the conclusion of the first year of research, we negotiated a contract to research and write a history of the firm situated in the context of regulatory development. This agreement assured us independent status within the corporation. We received full access to corporate records and personnel, but we agreed not to publish proprietary information such as documents relevant to on-going legal disputes. We received final editorial control of the manuscript, with the proviso that Panhandle Eastern could review drafts and make recommendations for our consideration. If such recommendations proved unacceptable to us, the company retained the right to insert footnotes explaining its views. We also received final disposition of the manuscript meaning that we intended to publish the completed work. Our goal was to research and write a university press quality book. The choice of a university press required that the final manuscript would undergo scrutiny by professional historians in a review process allowing us to set academic standards for our research. To avoid possible confusion about the focus of the book, we included in the contract a tentative chapter outline.

Beginning in the fall of 1992, Castaneda and Smith began full-time research, and Pratt served as project advisor. The company provided us with ample office space at its Houston headquarters and two computers and a printer. We also suggested that Panhandle Eastern form a reading committee to read and comment on drafts of the manuscript. This committee (see acknowledgments) was composed primarily of company retirees.

Our research covered a wide variety of sources both inside and outside the company. Panhandle Eastern stores nearly 50,000 boxes of corporate records in a large salt mine in Hutchinson, Kansas. A well maintained cataloguing system allowed us to ascertain fairly quickly which of those boxes would be useful for the study, and we ordered those which promised to be useful. In every case, we received requested materials within two to three days. We separated the firm's most valuable historical documents and photographs; they are now archived at the Woodson Research Center, Rice University.

Initially, we did not find much early executive correspondence or many internal documents regarding some of the company's short-lived diversification efforts. Further investigation uncovered a valuable collection of executive correspondence covering important issues from the 1930s through the 1950s; many of these materials had been stored outside of the corporate records management system. Eventually, we compiled a varied selection of diverse material covering nearly all of the firm's major activities.

Of the external documents we utilized, the most valuable were government documents. Since Panhandle Eastern required regulatory approval for major decisions involving its gas pipelines, the records of the Federal Power Commission and its successor, the Federal Energy Regulatory Commission, as well as those of the Securities and Exchange Commission, proved particularly useful. Perhaps the most unexpected find was an incredibly extensive array of correspondence, statistics, and testimony relating directly to Panhandle Eastern collected by the Federal Trade Commission during its intensive investigation of the utility industry between 1928 and 1935.

We also used a wide variety of other sources. An extensive newspaper clipping file dating from the mid-1930s forward provided useful information as did collections of internal reports and studies. Several employees and retirees sent pertinent records on particular topics to us after the company advertised our existence – and our interest in historical materials – in the company employee newspaper, *Pipelines*. We also reviewed the usual assortment of minutes, annual reports, and various company publications as well as secondary sources and articles from newspapers as well as technical and trade journals. A surprisingly useful source included many published and unpublished histories of other gas and utility firms. Oral history interviews of a wide range of current and former employees provided insight into aspects of the firm's history not found in archived records and informed us of the firm's cultural and social history.

Acknowledgments

There are many persons who deserve credit for participating in this effort. Panhandle Eastern CEO Dennis Hendrix's interest in preserving the history of the natural gas industry initiated the impulse for the research and writing of this work. Soon after Hendrix joined Panhandle as its CEO and chairman in November, 1990, Jim Hart invited Pratt and Castaneda to submit a proposal to commence a history project at Panhandle Eastern Corporation. Hart worked diligently in developing a cordial and open atmosphere for our project and facilitated the process of gaining approval of our proposals. His active role in protecting the integrity of our mission, contract, and overall project allowed Castaneda and Smith the freedom necessary to successfully complete this project in a timely manner.

We express great thanks to Louis Galambos, Professor of History at The Johns Hopkins University, who prodigiously edited the manuscript and in so doing clarified significant economic, regulatory, and entrepreneurial issues while at the same time making the manuscript substantially more enjoyable to read. Frank Smith, executive editor at Cambridge University Press, smoothed the path through the publication process, and we owe him our great thanks as well. Robert Lewis of BHG assisted in all phases of contract negotiation and provided valuable administrative assistance.

We also received a good deal of useful input from the members of a reading committee which received draft copies of manuscript chapters. Reading committee chairman Mack Price periodically consolidated all of the reading committee's comments onto one manuscript and passed it back to us for our consideration. Price assisted us in compiling a list of persons to interview, located historical records, and catalogued a long series of journal articles about the company. Other members of the reading committee included Robert D. Hunsucker, Thomas B. Irwin, John A. Irwin, Arnold J. Levin, John F. Schomaker, Robert A. VanLeuvan, Raymond M. Shibley, and Tony Turbeville. Additional readers were Dorothy Ables, James Castaneda, Dennis Hendrix, Max Lents, Trey Mecom, Richard L. O'Shields, Joseph Pratt, and John Sieger. Two anonymous readers for the Press provided particularly helpful comments and suggestions.

We acknowledge the posthumous contributions of Professor Robert B. Eckels, a former professor of history at Purdue University, who worked on a history of Panhandle Eastern during the 1950s and early 1960s. This earlier effort resulted in a draft manuscript, and the organization of a great deal of significant historical information into approximately thirty boxes. Eckles' work was never published (see Appendix I for a brief review of this earlier project), but it was a valuable source for our research.

Numerous other persons assisted us in a variety of ways with our work, and we apologize to those we fail to mention. Tony Turbeville served as our immediate in-house contact and proved very adept at directing us to people or material useful to our study. Tony also helped coordinate the reading committee, locate photographs (along with Tom Overton), and assisted us in innumerable ways. Tan Nyguyen and Pam Horsman created the maps and charts used in the book. Elizabeth Edrich prepared the index. Others who provided a wide variety of assistance include John Barnett, Maureen Civiletto, Dan Dilks, Elizabeth Easton, Martha Grimmet, Leona Harrison (for the previous Texas Eastern history), Jan Huber, Donna Lucas, Trey Mecom, Mabel Menefee, David Munday, Diane Nesrsta, Bob Reed, Keith Schmidt, Shari Stafford, and Carol Sullivan. At Interstate Natural Gas Association of America, Skip Horvath, Josephine Hacey, and Samir Salama provided us with valuable statistics for use in Chapter 10. The history departments of both the University of Houston and California State University, Sacramento provided collegiality during the course of this work. Terri Castaneda and Vicki Flemmer typed the final version of the manuscript. Candace Morton at Huron Valley Graphics guided the manuscript through production. Despite much help from all of these persons, the authors claim responsibility for all errors.

Finally, we thank our respective families, Terri, Courtney, and Ramsey Castaneda and Denise Kettelberger, Elizabeth and David Smith for their patience, encouragement, and support during the course of this project.

Gas pipelines and the emergence of America's regulatory state

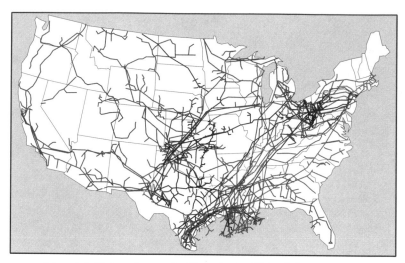

Map. 1.1 Natural gas pipelines in the United States, 1993

1

Introduction

At a spring 1991 meeting of his pipeline company managers, Paul Anderson, then Panhandle Eastern Corporation's group vice-president, wrote the phrase "merchant service" on the blackboard, sketched a circle around the words, and drew a diagonal line across them. "We are going to assume," he announced, "the merchant service is dead. We are going to go forward on that basis."[1] As Anderson's comments indicated, the pipeline firm had accepted the fact that an entirely new regulatory policy had emerged; he was preparing a strategy that would enable the company to operate successfully under the new rules.

Within a year, the Federal Energy Regulatory Commission (FERC) issued Order 636, which chairman Martin L. Allday said ". . . will signal the end of the seemingly endless transition period the gas industry has been in. . . . [Order 636] is the next and hopefully last major step in the Commission's efforts to allow competition rather than regulation to govern how pipelines function."[2] Responding to years of failed regulatory policy, the FERC installed a new approach for the U.S. pipeline industry (Map 1.1) that involved replacing the traditional mer-

[1] Paul M. Anderson, interview by Christopher J. Castaneda and Clarance M. Smith, December 21, 1993, PECA.
[2] See FERC, *Fact Sheet*, Remarks of FERC Chairman Allday, April 8, 1992.

chant function – the purchase and sale of natural gas – with a modified common carrier status.

Much earlier in the twentieth century, pipelines had been important parts of large, vertically integrated utility companies which in most cases were in the business of discovering, producing, transporting, and distributing gas as well as electricity. Regulatory reforms of the 1930s broke apart these large combines and allowed merchant oriented pipeline firms to exist in smaller systems which were more commonly integrated with a gas production subsidiary rather than a local gas distribution firm. Beginning in the 1950s, increasingly restrictive regulations prompted many pipeline firms to begin diversifying into a wide range of related (as in oil production) and unrelated (real estate development) activities. By the 1970s and 1980s, numerous pipeline systems were the core business of diversified energy firms struggling to respond to rapidly changing market conditions and economic conditions. Order 636 was a decisive regulatory answer to uncertainty in the gas pipeline business as it transformed interstate pipeline companies into "nondiscriminatory, open access transportation systems," or contract carriers (pipelines which transport, but do not buy and sell, gas for producers and/or consumers). While this new policy does not prohibit vertical integration, it segments gas pipelines so that they compete only against other pipelines in the single economic dimension of transporting natural gas.

The Panhandle Eastern meeting in 1991 and the Commission's Order 636 presented two contrasting visions of the American political economy. These two concepts of how the economy should work, whom it should benefit, and what roles private and public decision making should play in its development had been central features of American political and business affairs for over a century. At Panhandle Eastern and elsewhere in American business, the primary ideology stressed entrepreneurship and the achievement of economic efficiency in a long-run process of capital formation. Where political authority intruded on business activity – as it had increasingly done in the past century – the government was one among other constraints that had an impact on competitive performance. Like markets, regulatory agencies had to be understood, their rulings anticipated, and managed so as to acquire competitive advantage in this realm as in others.

The contrasting visions embodied in the FERC Order 636

stressed equity and economic security, and this outlook featured an emphasis on existing stakeholders and the short-run development of the industry and its markets. The nation's regulatory state emerged in large part because many Americans became convinced that markets alone could not protect consumers and employees from the power exercised by business interests. For more than a century, the United States searched for a workable balance between regulatory control and a market-oriented, entrepreneurial brand of capitalism. Order 636 was merely one of the many recent efforts to change that balance, and this on-going process had done much to shape the nation's natural gas industry and Panhandle Eastern.

Evolution of a firm and a policy

Panhandle Eastern's evolution provides an informative historical perspective on the development of regulatory policy and the modern corporation. This firm's history traverses three significant eras in the growth of gas pipelines and the complex interactions of entrepreneurs, managers, regulators, technicians, and judges in that process. The first era, in which the foundation of the new industry was laid, was characterized by intense business development and competition with minimal regulation or antitrust activity. The second was the era of the regulated merchant function – buying and selling gas, in which pipelines enjoyed a long period of expansion in spite of the increasingly complex regulations imposed on the firms. The third was characterized by regulatory failure, regulatory change, and a decisive reorganization of the industry as the market for corporate control – defined as those market transactions which result in a change of leadership and/or ownership of a corporation – developed in America and abroad.

Free competition: Captive markets

This history begins at a time characterized simultaneously by great strides in industrial innovation and, in the words of historian Richard Vietor, "the failure of competition."[3] In the mid-

---

[3] See, Richard H. K. Vietor, *Contrived Competition: Regulation and Deregulation in America* (Cambridge, Mass.: The Belknap Press of Harvard University Press, 1994), p. 2.

1920s, American business interests were relatively free of political controls. It was a time of rapid and unrestrained economic expansion coupled with fierce competition; yet society honored its entrepreneurs and their business vision. The fruits of economic expansion were of great value to Americans, who appreciated the jobs, new products, and services they received from several industries, including gas and electric, which were growing extremely fast and extending their influence into new markets throughout the country.

The "roaring twenties" witnessed an extended period of peace and prosperity in the United States. Innovations in electric power generation and transmission were reshaping industry and urban life, while improvements in the manufacture of airplane and truck engines, as well as tires, gave these forms of transportation new advantages over older industries like railroads. In the natural gas industry, technological and market developments were beginning to propel industry growth.[4]

Until the mid-1920s, the natural gas industry remained undeveloped. Existing pipeline technology could not prevent pipelines of significant length from leaking; prior to 1925, only one line reached 183 miles and this one was the exception. Industry development was thereby restricted to parts of the Appalachian and Gulf Coast regions where natural gas production was located near industrial and commercial customers. The majority of large energy consumers were not located near significant quantities of natural gas, and they relied on other sources of energy, typically coal. In many towns across the nation, manufactured coal gas plants produced a comparatively inefficient synthetic gas which could be piped locally. Manufactured gas provided lighting and heating for those who could afford it.

The development and application of electric arc welding in the mid-1920s made possible the construction of long-distance, leakproof pipelines. At that point, several entrepreneurs began to consider the possibility of building pipelines to transport gas from the huge and newly discovered fields in the southwest to existing manufactured gas markets in large midwestern urban areas. Thus, the discovery of massive gas fields in the Texas Panhandle during the first two decades of the twentieth century combined with improvements in pipeline technology created unusual opportunities for those willing to invest in this new industry. The result was a dramatic surge in long-distance pipeline construction during the late 1920s.[5]

An assortment of independent entrepreneurs and representatives of large public utility companies with secure access to metropolitan energy markets set out to connect energy hungry markets with newly available natural gas. During the late 1920s, they organized pipelines, including Panhandle Eastern, to buy natural gas from producers and sell it to their own local distribution companies or directly to industrial firms. These lines were, from the beginning, in the merchant–gas business as well as the transportation business.

There were no federal agencies or regulations which oversaw the natural gas industry. The Interstate Commerce Commission which regulated interstate railroad and oil pipeline rates was barred from doing the same for gas and electric firms. Only state regulatory agencies oversaw pipeline operations, and their influence was generally limited to setting the price of gas sold to local distribution companies for resale to residential and commercial customers within city or county jurisdiction. Pipelines which transported gas across state lines escaped for all practical purposes any regulatory oversight. Even intrastate lines could avoid state regulation simply by routing gas across the nearest state border and then bringing it back into the original state for sale. Some firms simply obfuscated regulatory officials by claiming that any interstate business done by their firm exempted all their operations from state regulation. In these years, state regulators simply could not regulate either the performance or the financial structure of interstate utilities like the pipelines.

Constrained only by market forces, gas pipelines rapidly expanded and became heavily involved in the process of combination that was taking place in most public utilities. Like other capital-intensive industries, pipelines were drawn by the carrot of economies of scale and repelled by the stick of intense competi-

---

[5] Ralph E. Davis, "Natural Gas Pipe Line Development During the Past Ten Years," *Natural Gas*, vol. 16, no. 12 (December 1935), pp. 3–8. Also see E. Holley Poe & Associates, *Development of Natural Gas Transportation in the United States* (E. Holley Poe & Associates, New York, 1946) and C. Emery Troxel, "Natural Gas Pipe Lines," *The Journal of Land & Public Utility Economics*, vol. 12, no. 4 (November 1936), pp. 344–54.

tion. Throughout the country, large gas and electric distribution systems and coal firms were being consolidated into multi-layered public utility holding company systems. Pipelines were drawn into these combines, which could take advantage of the growing national market for securities to raise the capital they needed. Panhandle Eastern, which had begun as an independent pipeline, was acquired – temporarily – by one of these giant utility holding companies. These combines had significant "first-mover" advantages; and captains of public utility firms were not hesitant to use them in an era in which enforcement of the antitrust laws was in abeyance.[6]

While the unregulated utility holding companies had done much to extend service rapidly to American consumers, they were politically vulnerable on two fronts. Many of them held virtual monopolies in large metropolitan areas and even in entire regions of the country. Americans were generally fearful of monopoly, especially where goods and services they considered to be necessities were concerned. This was particularly the case where utilities had become involved in illegal or suspect financial manipulations. Many utility companies were heavily leveraged, using debt rather than equity to finance their operations. That worked very well in a growing market, but when the great contraction began in the fall of 1929 the weaknesses of these leveraged holding companies became obvious. The collapse of the Insull System left little doubt in the public's mind that something had to be done to these businesses.[7]

Financial mismanagement in the utility industry, monopolistic practices, and suspiciously high energy costs became even more objectionable after the Great Depression which began in the early 1930s. In response, reform minded New Dealers insti-

---

[6] Alfred D. Chandler, Jr., *Scale and Scope: The Dynamics of Industrial Capitalism* (Cambridge, Mass.: The Belknap Press of Harvard University Press, 1990), p. 34. Also see, Jerome K. Kuykendall, "Antitrust Laws and Regulated Companies under the FPC," *Public Utilities Fortnightly*, March 17, 1960, pp. 373–81. Kuykendall, chairman of the Federal Power Commission in 1960 wrote: "utilities, including those subject to the Federal Power Commission, are not readily controlled by market forces and the antitrust laws are therefore not an apt means of regulation."

[7] See, Harold L. Platt, *Electric City: Energy and the Growth of the Chicago Area, 1880–1930* (Chicago: University of Chicago Press, 1991) and Forrest McDonald, *Insull* (Chicago: University of Chicago Press, 1962).

tuted an array of new regulatory agencies and controls designed to impose order on, and to ensure equity in, an economy
and society as yet unclear how to respond to the changes taking
place during those years. Seeking recovery and reform, the
country's political leaders reinvigorated existing business policies such as antitrust and created a number of new federal
agencies and laws to regulate firms made independent by new
antitrust enforcement policy. The antitrust laws had been
passed in an effort to ensure that American markets remain
competitive and free from domination by large, corporate monopolies. The application of this policy to utilities and natural
gas began in earnest after the passage of the Public Utility
Holding Company Act of 1935 (PUHCA). This Act allowed the
government to break apart the large public utility holding companies, and it separated many gas pipelines, local distribution
companies (LDCs), and producers.

Single industry regulation could now be imposed on the
utility industry. The Natural Gas Act (1938) gave the Federal
Power Commission (FPC) authority to regulate the interstate
natural gas industry, and the Federal Power Act (1935) gave
the same agency oversight of electric power. Meanwhile, the
Securities & Exchange Commission (SEC) acquired the authority to regulate the securities markets in which pipeline
companies and other firms acquired their capital. The subsequent history of Panhandle Eastern and the industry was
to a considerable extent a narrative of regulatory experimentation and entrepreneurial efforts to continue to develop
the industry under the new conditions these public controls
created.

New Deal antitrust policy had allowed Panhandle Eastern to
regain its independence from one of the largest public utility
holding companies, but it shed one form of central control only to
be subjected to another. Now the firm had to operate under the
cost-of-service regulation introduced by the Natural Gas Act.
This law did not eliminate competition in the industry; it merely
shifted it to a new arena. To be successful, entrepreneurs had to
be as adept at dealing with regulatory institutions as they were
at handling markets. By the early 1940s, the national regulatory system spawned by the New Deal was solidly in place (in
this industry and others), bringing new measures of order and
economic security to these markets. So long as the regulatory
state's basic economic context did not change, it would be able to

keep this "state cartel" functional.[8] Ultimately, that system too would be forced to give way, but for a number of years it appeared that Americans had been able to blend the best aspects of the entrepreneurial and the regulatory visions.

Managing regulation: The challenge

The second era of this history was highlighted for some years by post–World War II economic growth, prosperity, and stability. A variety of New Deal era regulations and regulatory agencies designed to promote stability – those for pipelines included – seemed to do just that. Market forces were so strong, it appears, and the nation's potential competitors so weak that even poorly conceived regulations could not hinder the progress of U.S. capitalism. Between 1939 and 1968, the financial, banking, oil, electric, airline, trucking, and telecommunication businesses all grew at a rapid rate. So, too, did the natural gas industry and the pipeline companies which now played a central role in the development of new markets for this fuel.

The rules set by the Natural Gas Act were not all particularly clear or concise, and the Federal Power Commission which administered it depended to a considerable extent upon the judicial system to define the regulatory context more clearly. For pipeline company managers such as Panhandle Eastern's William Maguire, this fluid situation offered many opportunities to achieve competitive advantage.[9] Maguire incessantly sought solutions to the problems posed by regulation, and he was as successful at regulatory competition as he was at market competition. While government regulators sought to stabilize the industry, Maguire sought innovation and economic change. To his credit, he made Panhandle Eastern a powerful force in the industry, while irritating his competitors, federal regulators, and several of his best customers.

[8] Jordan A. Schwarz, *The New Dealers: Power Politics in the Age of Roosevelt* (New York: Alfred A. Knopf, 1993), p. xi.

[9] See, Neil A. Fligstein, *The Transformation of Corporate Control* (Cambridge, Mass.: Harvard University Press, 1990), p. 2. Fligstein described well the behavior of managers such as Maguire in regulated industries when he wrote: "When one solution was blocked by the actions of the government, new solutions were created and diffused."

Panhandle Eastern had, for instance, tremendous natural gas reserves. In order to avoid federal regulation of these holdings, Maguire spun off a significant percentage of the reserves into a separate intrastate production company. He also sought to sell natural gas directly to industrial consumers, bypassing local distribution companies as well as FPC rate regulation. These efforts were vociferously opposed by Panhandle's largest customer, the Michigan–Consolidated Gas Company of Detroit, which did not want to lose its valuable industrial customers to Maguire's pipeline. This launched an extended regulatory – judicial struggle as to which company would be allowed to serve this market. These were the sorts of competitive struggles that characterized this middle era of Panhandle's history and the history of business–government relations in post–World War II America.

These struggles also help explain why the government inevitably sought to extend regulatory control. Cartels, state cartels included, are always fragile, always threatened from outside. The solution that usually seems logical to regulators is to extend control to cover the threat to stability. Originally, the Natural Gas Act of 1938 did not authorize the FPC to regulate gas production. The FPC only regulated the price at which pipelines sold gas. This gave pipelines opportunities to justify higher sales prices by controlling the transactions through which they purchased natural gas. Even more complicating was the fact that many pipeline firms owned significant reserves, again giving them leeway in setting prices for their gas.

Not surprisingly, the FPC sought to solve this problem by extending its control to gas production. In 1954, the U.S. Supreme Court's Phillips Decision mandated the regulatory agency to begin regulating the wellhead price of independently produced natural gas. While the case and the decision were complicated, the impact of this extension by the regulatory state was fairly easy to understand. It left the industry encased in a complex system of price controls and encouraged pipelines to begin diversifying into unregulated businesses. The FPC's stringent regulatory policies would become obsolete when the energy economy changed decisively as it did in the 1970s. While this outcome was predictable, the traumatic manner in which it would take place was not.

Search for a new equilibrium

America's long era of post–World War II growth and prosperity came to an end in the late 1960s. Domestic inflation started to rise and the rate of increase of the Gross National Product (GNP) began to decline. By the early 1970s, commercial banks started failing; airlines were losing money on once profitable routes; and natural gas and electricity shortages plagued northeastern cities. Mergers and acquisitions became the order of the day as many U.S. businesses underwent restructuring in an unpredictable economic environment. Global competition was already cutting into the markets of American firms, long before the nation began to question its military resolve in the wake of the Vietnam debacle.

According to economists such as Alfred Kahn and George Stigler, one of the major sources of U.S. economic malaise was a rigid regulatory system.[10] In natural gas, there could be no other conclusion.[11] The dramatic East Coast natural gas shortages were the direct result of the Phillips Decision which had allowed regulators to depress the price, and therefore supply, of interstate natural gas well below market requirements. Certainly, the 1973 Arab oil embargo aggravated the natural gas shortages because industry needed to use natural gas in place of embargoed oil. But sufficient quantities of natural gas were not available because producers had been given little profit incentive to find and develop new gas reserves dedicated for the interstate market. FPC regulations had simply deprived the natural gas pipeline industry of its ability to develop an adequate supply for the markets it had created. With the regulatory state in crisis, many sought a return to the entrepreneurial vision. If pipelines could be made competitive, most businessmen and academics argued, market forces rather than regulation would con-

---

[10] For a brief and concise overview of this era see Vietor, *Contrived Competition* pp. 12–14. Also see Alfred E. Kahn, *The Economics of Regulation: Principles and Institutions* 2 vols. (New York: Wiley); and George J. Stigler, "The Theory of Economic Regulation," *Bell Journal of Economics and Management Science* vol. 2, no. 1 (1971), pp. 3–21.

[11] Paul W. MacAvoy and Robert S. Pindyck, *The Economics of the Natural Gas Shortage (1960–1980)* (Amsterdam: North-Holland Publishing Company, 1975), pp. 16–17.

trol the industry and ensure that business and residential users would have the energy supply they needed.[12]

As regulators, politicians, consumer groups, and business debated the industry's structure, the pipeline companies, still responsible for supplying gas, sought alternative sources for their customers. Panhandle Eastern invested in all of the major sources of supplemental gas. Supplemental gas projects included coal and synthetic coal gas development; initiatives to import natural gas from Canada and Alaska; and the importation of liquefied natural gas (LNG) from North Africa. Despite the tremendous costs involved in these developmental projects, the economics appeared reasonable in the mid-1970s when some economists predicted that the price of oil would hit $100 per barrel in the 1990s.

Then, however, deregulation quickly eliminated the gas shortage, undercutting the very supplementary fuel ventures the prior regulations encouraged. The pipeline industry entered a period of organizational turmoil comparable to the late 1920s and 1930s. New combines emerged and diversified companies began to shed those functions not directly related, technologically and economically, to their core businesses. Against this background, the federal government launched one more experiment with regulatory control. It determined now that producers should compete only against other producers for gas sales, not against pipelines. Pipelines, as open access transporters, would compete only for the opportunity to transport gas contracted between producers and consumers. Order 636 determined that pipelines would operate somewhat like common carriers, barred from direct participation in the merchant service. It was this latest turn in the regulatory/entrepreneurial mix that prompted Paul Anderson to call his meeting and to encourage his managers to begin planning at once how they would keep the spirit of innovation alive under these new conditions. The story told in the pages that follow, the history of Panhandle Eastern and the regulatory state, will, we trust, help you understand these events and their meaning in America today.

[12] Stephen F. Williams, *The Natural Gas Revolution of 1985* (Washington, D.C.: American Enterprise Institute for Public Policy Research, 1985), p. 22. Also see Edward C. Gallick, *Competition in the Natural Gas Pipeline Industry: An Economic Policy Analysis* (Westport, CT: Praeger, 1993).

PART I

Free competition: Captive markets

2

Promoting Panhandle Eastern

Philip Gossler, the sixty-year-old chairman of Columbia Gas & Electric Company, was angry. Since he headed one of the largest public utility holding companies in the United States, he was capable of doing something about the businesses that made him angry: "If I would take a map of the Central United States and a pencil and a ruler," he told William Maguire in June, 1930, "I could not draw a line to raise more hell than that one."[1] That "line," extending from the Texas Panhandle to Indianapolis, was the pipeline Panhandle Eastern was proposing to build. Gossler was in no mood to sanction the construction of a maverick gas pipeline system positioned to serve midwestern markets – and he was not alone.

Gossler told Maguire what would happen if he persisted with a "line" that crossed into the firmly controlled markets of several powerful gas and electric firms (Map 2.1). The pipeline's path would take it near Kansas City, where Henry Doherty controlled the gas and electric power business through his Cities Service Company. The new line would also serve customers in

[1] Federal Trade Commission, Report to the Senate on Public Utility Corporations, Senate Document no. 92 (70th Cong., 1st session), part 82 (Washington, D.C.: GPO), p. 307. The hearings for this investigation began in 1928 and lasted into 1935. Future citations of this investigation will be in the format of 82 FTC Report 307.

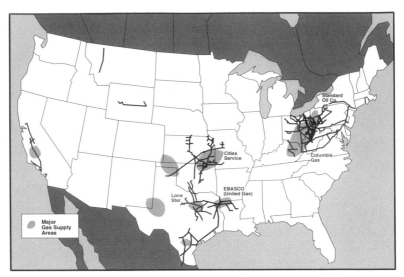

Map 2.1 Major natural gas pipeline systems, 1929

Peoria, which was utility "captain" Samuel Insull's electric and manufactured gas territory. It would pass near St. Louis, which was dominated by Christy Payne, who managed Standard Oil of New Jersey's natural gas operations. Worst of all, it would cross into Indiana on a path to Michigan and Ohio, Columbia Gas's domain. Gossler told Maguire "that it should be stopped; that it would influence the price of gas securities, and be generally demoralizing" to him as well as Mr. Insull. Gossler made clear that all these powerful firms would be injured and all would fight back.[2]

They could bring substantial resources to that battle. A small number of public utility holding companies controlled much of the gas and electric business in the United States. By owning controlling interests in the stock of other firms in the industry, these holding companies had built up large empires of operating companies. Two of the largest holding companies, Columbia Gas and Standard Oil of New Jersey, distributed more than half of the gas sold in the entire Appalachian region. Doherty's Cities Service dominated the lower midwest, but the largest of the public utility conglomerations was Middle West Utilities, Inc., and Insull Utility Investments, Inc., controlled

[2] Ibid.

by Samuel Insull and headquartered in Chicago.[3] Insull's empire included 248 gas, coal, and electric power firms serving 4,741 communities in thirty states. As Table 2.1 indicates, these four holding companies, along with Electric Bond & Share Company, controlled most of the gas and electric business in the midwest, south, and northeast and owned 58 percent of the nation's gas pipeline mileage.[4]

These large firms worked together in an informal, unstable, yet powerful cartel. Politicians labeled the public utility industry during this period as the Power Trust and lashed out at the concentration of economic power in the cartel's hands. As Pennsylvania governor Gifford Pinchot informed the Pennsylvania legislature, "nothing like this gigantic monopoly has ever appeared in the history of the world . . . if uncontrolled, it will be a plague without previous example. If effectively controlled in the public interest, it can be made incomparably the greatest material blessing in human history."[5] The "material blessings" were obvious. The highly capital-intensive gas and electric business was providing relatively inexpensive energy to more and more consumers. But as the cartel became a powerful economic and technological force in American society, many began to wonder if it could be "effectively controlled."[6]

---

[3] Insull and the development of gas and electric power have been the subject of numerous studies. See McDonald, *Insull*; Platt, *The Electric City*; and W. G. C. Rice, *Seventy-Five Years of Gas Service in Chicago* (Chicago: Peoples Gas Light & Coke Company, 1925).

[4] Carl D. Thompson, *Confessions of the Power Trust* (New York: E. P. Dutton and Co., Inc., 1932), pp. 232–3; James C. Bonbright and Gardiner C. Means, *The Holding Company* (New York: McGraw–Hill, 1932); Albert F. Dawson, *Columbia System: A History* (New York: J. J. Little and Ives Company, 1938), pp. 6, 159–71; Tussing and Barlow, *The Natural Gas Industry*, p. 32; and "Natural Gas," *Fortune* (August, 1940), p. 56; and 84-A FTC Report 36–45.

[5] Thompson, *Confessions of the Power Trust*, pp. xvii–xviii. Senators George Norris (NE) and Thomas Walsh (MT) were strong proponents of public power and initiated legislative attempts to impose governmental control over the public utility industry.

[6] See Thomas Parke Hughes, *Networks of Power: Electrification in Western Society, 1880–1930* (Baltimore: Johns Hopkins University Press, 1983). Hughes' thesis regarding the growth of the electric power industry rests on the concept of technological determinism. In the development of the twentieth century natural gas industry, it is evident that technological development opens markets but competition and regulatory policy shapes the implementation of technology.

Table 2.1. *Major public utility holding companies: Natural gas market control, 1930*[a]

Holding company	% of total natural gas production	% of total manufactured gas production	% of total natural gas distribution	% of total natural gas transmission	% of total pipeline mileage
Columbia Gas & Electric	4.2	0.7	16.4	12.2	24.7
Electric Bond & Share[b]	7.0	–	17.7	16.0	10.7
Cities Service Co.	3.7	0.4	19.3	24.9	14.0
Standard Oil (N.J.)	1.5	–	11.1	10.0	8.8
Total	16.4	1.1	64.5	63.1	58.2

[a]These four firms are the top four in each category listed above except for manufactured gas production. Insull's utility firms are not listed because they were engaged primarily in the electric and manufactured gas distribution business.
[b]Includes United Gas Company system.

Source: Federal Trade Commission, Summary Report of the Federal Trade Commission to the Senate of the United States . . . on Economic, Financial and Corporate Phases of Holding and Operating Companies or Electric and Gas Utilities, vol. 68–9a (Washington, D.C.: GPO, 1934–1935), pp. 848–59.

Pipelines played a central role in the industry's evolution. New southwestern gas reserves were now available, as was an improved pipeline technology. These developments encouraged several of the holding companies to plan long-distance lines to transport natural gas to their existing local gas distribution subsidiaries. One project designed to transport natural gas from Texas to Insull's gas distribution system in Chicago was the "Chicago Line," a joint venture of Insull & Sons, Cities Service, Standard Oil of New Jersey, and the Texas Company (later Texaco). Construction began in July, 1930. At the same time, a consortium led by North American Light & Power Company, which owned gas and electric properties throughout the midwest, began building a line from Texas to Minneapolis/St. Paul via Omaha. Insull did not have to worry about competition between these two ventures because he controlled 40 percent of North American.

It was this imposing network of organizations that Panhandle Eastern Pipe Line was challenging. Still at the very early stages of construction, an independent facing opposition from large entrenched competitors, its chances for success were not good.[7] Philip Gossler made that very clear. Through the "gentleman's agreements" between the heads of the large public utility holding companies, the markets had been divided and the industry stabilized. While there were disagreements among Insull, Gossler, Payne, and Doherty, they all agreed on one thing: the Panhandle Eastern project represented a "raid" into their developed market territories.

For the most part, the cartel members had served their markets without drawing heavily on natural gas. They relied primarily on a synthetic gas, manufactured coal gas, produced from the distillation of coal. Since the public utility holding companies typically controlled the coal reserves and coal gas plants as well as the local distribution systems, they were not eager to introduce natural gas, a substantially more efficient and less expensive fuel, to their markets. They had too much capital invested in the coal plants to discard them quickly. Instead they wanted to phase in the new product. They were willing to build pipelines which would introduce natural gas to cities where they already controlled the distribution systems and could firmly control the transition from manufactured to

[7] See Stotz and Jamison, *History of the Gas Industry*, pp. 381–5.

natural gas. But they were adamantly opposed to any independent firm controlling the gas flow and thus the transition.[8]

After William Maguire discussed this conflict with Philip Gossler during the summer of 1930, Maguire reported back to Frank P. Parish. Parish was the organizer of Panhandle Eastern, and he had engaged Maguire as a utility consultant a few months earlier to sign gas sales contracts with industrial and utility firms in Missouri and Illinois. It was not easy to sell the gas they planned to bring from Panhandle's Texas reserves to midwestern customers. Their formidable opponents had influence with the large New York financial institutions, with pipe manufacturing and supply firms, and with state public utility commissions. Without the financial and political support of at least one of the utility combines, an independent such as Panhandle Eastern had an uncertain future. Parish nevertheless was convinced that he could win this struggle and join the ranks of the magnates himself.

Shippey, Maddin, and Parish Partnership

Frank Parish was an unlikely challenger of the utility titans. At thirty-four years of age, the fair-haired, easy-smiling man was youthful in appearance even though he was already a seasoned salesman. His background was a mixed bag that included alternating episodes of mediocre employment, success and bankruptcy. After dropping out of junior high school in Sidell, Illinois, Parish caught a westbound train to California where he worked as a farm laborer and teamster. During the next several years, he found jobs as a store clerk in New York City, a machinist in Detroit, and a harvest hand in Saskatchewan, Canada.[9]

In 1913, Parish moved to Chicago and started his own business as a machine equipment broker. With the knowledge gained from this venture, he formed the Parish Supply & Manufacturing Company and began producing pulleys and tackle blocks. Parish and his brother, Lawrence, secured a gov-

---

[8] For an interesting theoretical discussion of economic transitions – and a good model for understanding the transition from manufactured gas to natural gas – see the theory of "creative destruction" in Joseph A. Schumpeter, *Capitalism, Socialism, and Democracy* (New York: Harper and Row, 1975).

[9] "Parish Tells His Meteoric Rise into Big Money," *Chicago Daily Tribune*, April 25, 1935.

ernment contract during World War I for 9,900 tackle blocks priced at about $35 each. Parish immediately sublet the contract and pocketed a profit of $60,000. After the war, however, demand for his products evaporated, and following the sudden inflation and deflation of 1920–1, the business failed as did many others in those years. The Parish brothers found temporary employment transporting peaches by barge to Chicago, then sold electrical equipment and tried the real estate business. In 1926, with $100,000 in debts, Frank Parish filed for personal bankruptcy.[10]

The following year, his fortunes improved. He accepted a sales job in Chicago with the pipe manufacturing firm of Clayton, Mark & Company, covering much of the central United States, including Kansas City.[11] One of his major customers was Cities Service Company.[12] At the time, Cities Service was the sole supplier of natural gas to Kansas City, but a group of investors had recently proposed to build a competing gas line from north Texas to the city. Eager for a second gas supply, the *Kansas City Star* attacked Doherty and his firm for the high prices stemming from its monopoly. "Cities served by the Doherty gas monopoly have been helpless," the paper charged. "A Doherty pipeline company, unregulated because of lack of laws, is the secret of the success of the system."[13] One editorial said:

> Suppose a man from Mars should visit Kansas City. Suppose he should inquire about the city's natural resources and finally come to the question of fuel, of gas. . . . "Well," he would be told, "the fact is that we aren't very bright. We have given a monopoly, yes given it, the privilege of selling us the gas at whatever price [it] chooses."[14]

---

[10] Ibid.

[11] For a useful study of the development of gas and electric utilization in Kansas City see Mark H. Rose, *Cities of Light and Heat: Domesticating Gas and Electricity in Urban America* (University Park: Pennsylvania State University Press, 1995), pp. 160–3 and Mark H. Rose and John G. Clark, "Light, Heat, and Power: Energy Choices in Kansas City, Wichita, and Denver, 1900–1935," *Journal of Urban History*, vol. 5, no. 3 (May, 1979), pp. 340–64.

[12] For a published history of Cities Service, see William Donohue Ellis, *On the Oil Lands with Cities Service* (Cities Service Oil and Gas Corporation, 1983).

[13] *Kansas City Star*, January 20, 1929.

[14] Editorial, *Kansas City Star*, July 30, 1927.

To counter this editorial campaign, Doherty purchased a substantial interest in the Kansas City *Journal–Post* and made it a friendly forum for his views.[15] Meanwhile, he began searching for additional low-cost gas so that he would, if necessary, be able to undercut the threatened competition and perhaps even appease disgruntled locals. Just then, pipe salesman Parish made a timely visit to the offices of the Gas Service Company, a subsidiary of Cities Service. There he learned from employee William J. Hinchey that two men named Shippey and Maddin needed money and pipe in order to build a line from a nearby gas field. If the line could be built, Cities Service would purchase the gas.[16]

C. Stuart Shippey was a first-class wildcatter who had invested considerable energy in prospecting the region near Belton, Missouri. After he had leased potential gas-producing acreage, Shippey joined forces with Samuel J. Maddin, a self-taught field engineer who had experience laying pipe and drilling wells.[17] Sam Maddin and Stuart Shippey believed they could produce gas and market it locally. Their acreage was near Belton, a community about forty miles south of Kansas City, where wildcatters had discovered oil during the late nineteenth century. They were convinced this oil field could produce shale gas if they drilled deeper than the wildcatters, and they thought they could sell the gas to the 7,000 or so homes located in the area between Kansas City and Belton, which had no gas service. At this point they had a supply and a demand – all they lacked was capital.[18]

This provided Parish with the opening he needed. After locating Maddin and Shippey in Belton, Parish quickly struck a deal. The three men formed a partnership in which each would

---

[15] *History and Information Concerning the Gas Service Company* (Gas Service Company, no date), p. 63.

[16] 82 FTC Report 167. Also see *History and Information Concerning the Gas Service Company, 1925–1957*, pp. 24–5, 36–7.

[17] Robert B. Eckles, untitled manuscript on the history of Panhandle Eastern, ca. 1959, Part 1, Chapter II, page 1, located at the Indiana Historical Society. Hereinafter, references to the Eckles manuscript will be cited in the form of Eckles, Pt. 1, C. II, p. 1, IHS.

[18] Luke J. Scheer, *Panhandle Eastern – The First Forty Years* (unpublished manuscript), pp. 50–61, located in the Panhandle Eastern Corporation Archives, hereinafter referred to as PECA.

have a one-third interest and Parish would serve as general manager. On November 1, 1927, they formally organized the Shippey, Maddin, and Parish Partnership, a business alliance which would last for several years, amass large amounts of capital, and provide immense profits for each of the partners. The handsome, young salesman and the two older, experienced field men made a strange combination, but it was one with substantial business potential. They drew into their new enterprise other knowledgeable gas men, including Hinchey who resigned from Cities Service to work for the partnership.[19]

Parish's first effort on behalf of the tiny organization was to negotiate a contract to sell shale gas to American Pipe Line Company, a subsidiary of Cities Service, which would purchase approximately 80 percent of the group's gas during the next several years.[20] The utility firm was pleased to do business with Parish as long as his efforts did not threaten its lock on the Kansas City market. Cities Service's support was crucial. It lent the partnership $40,000, and with the new contract signed, the partners were able to get a $250,000 line of credit for equipment and supplies from the National Supply Company of Toledo, a pipe distributor for U.S. Steel. This enabled the partners to begin laying the forty-eight-mile, twelve-inch line from the Belton gas field to the American Pipe Line's Holmes and 97th Street compressor station in Kansas City.[21]

In order to avoid state regulation by the Missouri Public Service Commission, American Pipeline routed the partnership's gas two miles west to the Kansas state line, where it passed through one of Cities Service's Kansas compressor stations before returning to Missouri for distribution within the city limits of Kansas City, Missouri. The circuitous route was lengthy and expensive, but it allowed Cities Service the advantage of using unregulated interstate gas instead of the price-controlled, intrastate product. Since no federal agencies or laws

---

[19] Frank P. Parish, interview by Robert B. Eckles, July 7, 1957, PECA.

[20] 84-A FTC Report 261. Cities Service agreed to purchase all of the partnership's excess gas available up to 12 million cubic feet (MMcf/d) for a period of five years. The specified rate was $.20 per thousand cubic feet (Mcf) for the first 5 Mmcf/d and then $.13 per Mcf for the remaining 7 Mmcf/d. Local utility companies across the state paid approximately $.40 per Mcf for manufactured gas.

[21] 82 FTC Report 168–9.

oversaw the operations of interstate gas lines, pipeline owners found that if they qualified their business as interstate they could exempt their operations, and specifically their rate structure, from local or state control.[22]

With this first substantial contract in hand, Parish was able to turn his attention to developing additional markets. The Jackson County Power and Light Company of Independence, Missouri, located five miles east of Kansas City, was the partnership's next customer. Cities Service did not oppose the contract. Instead, it simply purchased the Jackson County firm and extended its own control over regional gas distribution.[23]

By this time, Shippey, Maddin, and Parish had realized that there was a huge potential market for natural gas in the midwest. But before they could tap that market they needed capital for further exploration and additional production; if they had the money, they could acquire distribution systems and construct new transmission lines. Because Cities Service officials wanted Shippey, Maddin, and Parish to become one of their major gas suppliers, they introduced Parish to the underwriting firm of P. W. Chapman & Company to assist in raising capital. The partnership calculated that it had an immediate need for $1.5 million, but it would not be long before that sum would prove insufficient to satisfy their plans to expand the business.[24]

Missouri–Kansas Pipe Line Company

Shippey, Maddin, and Parish recognized the necessity of transforming their partnership into a corporation if they were going to raise the capital they needed for expansion. They formed the Missouri–Kansas Pipe Line Company, or Mo–Kan, on May 5, 1928. The first office, in Belton, Missouri, was located in the front of a long corrugated building, the back of which was used to store pipe and other supplies.[25] Parish, Mo–Kan's president, was responsible for both finances and the task of promoting the new firm. Shippey, a vice-president, was in charge of land and

---

[22] Ibid.

[23] Eckles, Pt. I, C. II, p. 2, IHS.

[24] 84-1 FTC Report 154–7.

[25] *Mo–Kan Bulletin*, vol. 1, no. 1 (August 1929), p. 1, PECA.

Figure 2.1 Frank Parish and Mo–Kan executives, 1930

lease acquisition, while Maddin, also a vice-president, oversaw drilling and pipe laying work.

The three founders quickly appointed a board of directors, the first step in persuading prominent persons to invest in Mo–Kan.

Along with Parish, Shippey, and Maddin, the original board included Charles A. Meyer, an executive of National Supply Company, the sole distributor for U.S. Steel's pipe manufacturing subsidiary, and E. G. Peterson. Other directors who joined the board within the year included Ralph P. Mayo, a Denver-based financier; Francis J. McManmon, a Chicago public utility expert; Burt R. Bay, a former Cities Service engineer who also became Mo–Kan's general manager and vice-president; and Ralph G. Crandall, a Mo–Kan attorney.[26]

On May 29, 1928, the board of directors convened for the first time. They had two tasks to perform: authorizing the purchase of the Shippey, Maddin, and Parish partnership assets and raising funds for expansion. The board approved the sale of bonds and stock, the proceeds from which were first to be applied to the purchase from the partnership. Mo–Kan paid the partnership $630,000 in cash – for assets which had originally cost only $247,000 – and then valued its new assets at $1,735,000, based on figures supplied by the engineering firm of Brokaw, Dixon, Garner, & McKee. The partnership made a substantial profit on the sale of its assets to Mo–Kan, a capital gain based on their optimistic predictions of what Mo–Kan could achieve and the concurrence of all those who bought the new firm's stocks and bonds.[27]

During the inaugural board meeting, the directors also approved a contract to construct an additional forty-seven miles of pipeline into Kansas City. On July 16, 1928, the directors

---

[26] Eckles, Pt. I, C. II, p. 7, IHS.

[27] See, 84-1 FTC Report 155, 82 FTC Report 170–2, and 84-A FPC Report 320–1. The board approved the sale of $1.5 million in 6½ percent gold bonds, issued on June 1, 1928 and due on June 1, 1940, underwritten by P. W. Chapman & Company; the bond sales netted $1.346 million. An additional offering of $500,000 in notes was issued in December. Stock purchase warrants were attached to the bonds giving each holder of $1,000 in principal amount the right to purchase ten shares of common stock at $5 per share before June 1, 1931. All bonds were sold for $90 each by December 4, 1928. Chapman & Co. also purchased $500,000 of one-year 6 percent convertible gold notes yielding Mo–Kan $470,000. The Mo–Kan board also sold 200,000 shares of stock at $5.00 per share; 15,000 shares went to the underwriters and the remaining shares were distributed among the three partners for future sale. Using proceeds from the bond and stock sales, Mo–Kan purchased the partnership's assets. See, Eckles, Pt. I, C. III, p. 4; and Eckles, Pt. I, C. II, p. 6, IHS.

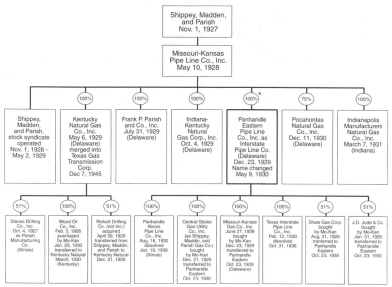

Chart 2.1 Organization of Mo–Kan interests

announced that the Shippey, Maddin, and Parish Partnership would receive $200,000 for the construction work and 2,000 shares of Mo–Kan preferred stock at $100 per share. By serving as general contractors for their own construction project – a standard practice of that time – the three men profited from significant contractor's fees.[28]

Frank Parish and his partners, confident in their new found ability to raise money, turned to other, bolder projects. Between 1928 and 1930, as Chart 2.1 indicates, Mo–Kan created, purchased, or otherwise achieved control of sixteen subsidiaries as it attempted to become a vertically integrated public utility holding company in the image of Cities Service or Columbia Gas. In one instance, Sam Shippey recommended that the partnership purchase Wood Oil Company, a small production firm which owned substantial oil and gas acreage in western Kentucky. This acquisition was intended as a signal to the stock-buying public that the firm was adding substantial reserves

---

[28] Transcript of Robert Eckles reading (June 6, 1956) from the Missouri–Kansas Pipe Line Company, *Minutes*, PECA.

and would soon be doing an even larger business. Eventually, Missouri–Kansas constructed both a small gathering system and a pipeline to transport this gas to Terre Haute, Indiana.[29]

These developments did not go unnoticed in the cartel. Like all such restrictive arrangements, the natural gas combine constantly beat back challenges to its control of the market. Originally Cities Service had supported the new pipeline, planning to benefit from the new supplies of gas and hoping to keep their small venture in line. But Mo–Kan's leaders were no longer content only to sell natural gas to Cities Service and to perform functions that had this large customer's approval (Map 2.2). Cities Service and its powerful leader, Henry Doherty, now turned against the upstart and made it extremely difficult for Mo–Kan either to expand further or to raise the funds it needed to pay the interest due on its bonds and notes.

Parish was ready to deal with this challenge, using some of the creative financial techniques common during the prosperous 1920s. The 1920s was a decade of experimentation – some of which was later outlawed – in the financial markets, and one of the popular devices was the stock syndicate. The original members of this syndicate were Parish, Shippey, and Maddin (who put their collective 185,000 shares of Mo–Kan into a pool), joined by former Cities Service employees, J. L. Clark, J. D. Creveling, and William J. Hinchey.[30] Parish used the syndicate to inflate the value of the stock and to persuade the firm's creditors to exchange their bonds and notes for common stock. Parish recalled that "the alternative [to defaulting] was to devise a means whereby holders of those obligations would exchange their bonds and notes for common stock."[31] This he did with the syndicate.

Syndicates included a limited number of participants who purchased large blocks of shares of a single corporation. A

---

[29] L. L. Waters, *Energy to Move . . . Texas Gas Transmission Corporation* (Texas Gas Transmission Corporation, 1985), pp. 15–29. These properties were later merged into the Kentucky Natural Gas Company. The three men also acquired property in their own names which was later sold to Mo–Kan at substantial profits.

[30] 84-1 FTC Report 155.

[31] Eckles, Pt. I, C. III, p. 5, IHS. For an interesting account of Wall Street see John Brooks, *Once in Golconda: A True Drama of Wall Street* (London: Gollancz, 1970).

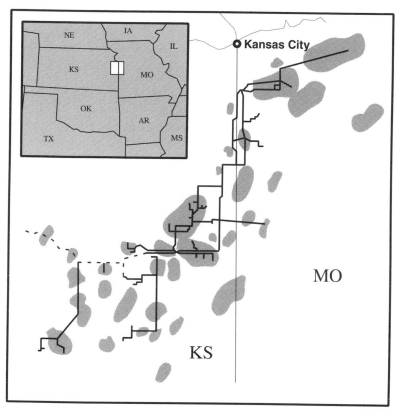

Map 2.2 Missouri–Kansas Pipe Line Company & gas fields, 1929

stock pool manager, often a specialist or even a trader on the
stock exchange, coordinated the participants' transactions.
Once the group took possession of the desired shares, the members traded them back and forth among themselves, gradually
accelerating the pace of trading in order to attract attention to
the stock. Investors watching the ticker-tape witnessed the
stock's heavy activity, and as the price climbed, the syndicate
members began to sell shares to unwitting investors.[32] Using
these techniques, Parish's stock syndicate performed well. In
February 1929, Mo–Kan converted all of its bonds and notes to
stock. During the next year, Mo–Kan financed its operations

[32] 84-1 FTC Report 155 and 84-A FTC Report 438–40.

Table 2.2. *Total amount of Mo–Kan stock underwriting,*
1929–1930

	Class A common stock	
Dec. 2, 1929	90,000 shares @ $15 per share	$1,350,000.00
Jan. 16, 1930	164,448 shares @ $20 per share	3,288,960.00
Apr. 3, 1930	400,000 shares @ $24.04 per share (Per agreement with Kentucky Natural)	9,614,732.38
Apr. 16, 1930	500,000 shares @ $28 per share	14,000,000.00
May 13, 1930	500,000 shares @ $29 per share	14,500,000.00
May 29, 1930	500,000 shares @ $33.625 share	16,812,500.00
	Total of 2,154,488 shares	$59,566,192.38
	(Average price per share = $27.648)	

Class B common voting trust certificates – April 15, 1930		
Apr. 15, 1930	818,351 shares @ $1.00 per share	$818,351.00
Total Liability on Underwritings		$60,384,543.38

Source: Scheer, Panhandle Eastern manuscript, Part I, Chapter III, p. 15.

solely through the syndicate's sale of common stock (Table 2.2).
Meanwhile, the syndicate's members reaped a $2.4 million
profit.[33]

One measure of the success of a stock syndicate was its ability to garner the participation of individuals outside the group,
especially news reporters and brokers. These men were important because they could plant stories in newspapers and offer
tips to clients. The ultimate achievement in stock syndicating
was to obtain the support of a prominent stock speculator.

Frank Parish was able to attract the attention of just such an
investor in Francis I. du Pont, a great-grandson of Eleuthere

[33] 84-1 FTC Report 155–7. Mo–Kan issued 77,090 shares of stock and 225
shares of scrip (each worth ¼ share of stock) to redeem $1,468,500 of the
First Mortgage Bonds. By the end of February, all of the bonds and notes had
been exchanged for Mo–Kan common stock; 52½ shares were exchanged for
each bond and fifty shares for each note. See, Eckles, Pt. I, C. III, p. 5, IHS.

du Pont who had founded the family's Delaware-based gun powder firm in the early nineteenth century. Francis managed DuPont's research laboratories and during his career laid claim to over a hundred patents.[34] In 1929, Francis du Pont purchased several thousand shares of Mo–Kan stock through Bioren & Company of Philadelphia. After learning of the purchase, Parish contacted du Pont hoping to gain further support from him and his family. As it turned out, Francis was the only family member interested in the venture, but he began working closely with Parish on the plans to expand Mo–Kan.

Parish clearly understood the value of the du Pont name. He and his colleagues elected the Delaware capitalist to the Mo–Kan Board in January, 1930, perhaps hoping that the public would assume the new company had the general backing of the distinguished family of East Coast industrialists. Parish, who made no secret of his ambition to become a famous entrepreneur, courted du Pont and included him in his plans for Mo–Kan's future. Parish, du Pont, and Crandall controlled a new voting trust that held five million shares of Class B common, authorized to elect four of the nine Mo–Kan directors. Since the three Class B stockholders also controlled more than 21 percent of the outstanding common stock, they could elect one additional director, giving them a majority on the board.[35]

Practices such as these had become normal during "the greatest bull market ever known."[36] Then the American public was anxious for more profitable investments than bank deposits, life insurance policies, or real estate. Stocks and bonds provided substantially higher rates of return than those of bank deposits and they were far more liquid than real estate.[37] At the end of 1928 – the best year ever for stock speculation –

---

[34] See Alfred D. Chandler and Stephen Salsbury, *Pierre S. Du Pont and the Making of the Modern Corporation* (New York: Harper and Row, 1971) and Daniel Gross, "Ross Perot's $100-million disaster," *Audacity* (Summer 1994) p. 7. Francis du Pont later founded the F. I. Du Pont & Company brokerage firm.

[35] Eckles, Pt. I, C. III, p. 9, IHS. The Mo–Kan board of directors voted to issue the Class B common stock in February 1930.

[36] 84-1 FTC Report 155. Also see Robert Sobel, *The Great Bull Market: Wall Street in the 1920s* (New York: W. W. Norton and Co., Inc., 1968).

[37] See Harold Cleveland and Thomas Huertas, *Citibank, 1812–1970* (Cambridge, Mass.: Harvard University Press), pp. 135–46.

Americans had become enamored of the market. "The market is now its own law," reported the New York *World*. "The forces behind its advance are irresistible."[38]

Along with this "irresistible" advance came abuses of power and highly questionable Wall Street marketing techniques. Federal and state authorities tolerated these abuses, and Wall Street operated throughout the 1920s largely unhindered by regulation. William Z. Ripley, a Harvard economist, had become worried about these developments.[39] He visited the White House in 1927 to discuss these problems with President Calvin Coolidge. Coolidge listened stoically as Ripley criticized corporate practices and explained that businesses often operated contrary to the public interest and that there was a detrimental concentration of economic power in a small number of firms. Impressed with the logic of the professor's argument, Coolidge removed his cigar and asked, "Is there anything we can do down here?" Ripley answered that the President was unable to act under current law – state action, he said, was needed to solve these problems. Cool Cal relaxed and sighed, comfortable in the knowledge that he could not be blamed for economic problems beyond his ability to control.[40]

Parish, the stock promoter, thrived in this environment. Recognizing his ability to obtain capital in Chicago's financial district, the Chicago press referred to him as the "boy wizard of LaSalle Street." Indeed, Parish had quickly acquired the trappings of success. Soon after profiting from the sale of the partnership's gas properties to Mo–Kan, he purchased a $100,000 home in a fashionable section of Kansas City. Later, he bought the former presidential yacht, the *Mayflower*, from the United States government.[41]

He was not alone in riding the stock boom of the twenties to

---

[38] *New York World*, no date, PECA.

[39] William Zebina Ripley authored several books including the following: *Main Street and Wall Street* (Boston: Little, Brown, and Company, 1927); and *Trusts, Pools and Corporations* (Boston: Ginn and Co., 1905).

[40] Robert Sobel, *The Big Board: A History of the New York Stock Market* (New York: The Free Press), p. 236.

[41] *Kansas City Times*, November 25, 1931 and Frank Parish, Jr., telephone interview by Clarance M. Smith, October 22, 1992. Parish purchased the yacht for $16,105 after it was razed by fire in Philadelphia. Parish already owned a 150-foot vessel named *Theo* in honor of his wife.

a new position in society. Utility stocks highlighted the glamour of Wall Street during these years. Samuel Insull's huge public utility holding company empire dominated the industry; by 1930, more than one million persons owned stock in the Insull industries. Halsey, Stuart & Company, an investment banking firm located in Chicago, helped by pumping up public interest in the Chicago utility.[42] For many years, electric power firms had provided stock markets with the leading American utility securities. Natural gas men in the 1920s saw signs that their industry now had a chance to enjoy the same tremendous growth in investor interest.[43]

In this setting, Parish took every opportunity to increase the activity in Mo–Kan stock sales. On May 2, 1929, the Parish stock syndicate sold its remaining 60,000 shares to a professional brokerage firm, Pynchon & Company, for $30 per share. After Pynchon & Company (apparently apprehensive about its large holdings in Mo–Kan) asked Parish to take back the task of selling Mo–Kan stock, Parish formed Frank P. Parish & Company (July 1929) to become the sole broker of the shares. Throughout 1929 and the first half of 1930, Frank Parish & Company sold and bought Mo–Kan stock. Parish opened headquarters in New York and branch offices in Illinois, Missouri, Wisconsin, and California. He hired Edgar H. Stapper to run the company, and Stapper worked through an extensive network of brokers and salesmen to create a market for new stock. Backed up by an advertising campaign, Stapper aggressively sold the stock, while purchasing as many of the shares as he could from other brokers when offered at below market prices.[44]

Mo–Kan stock traded as high as $35 per share during August, 1929, but it quickly dropped to $8 per share following the Wall Street crash. During the next year, Parish & Co. managed to "peg" Mo–Kan stock (Table 2.3), keeping its value up by repurchasing all stock offered for sale. Self price supporting was common in the years prior to the creation of the Securities and Exchange Commission (SEC). From August 1, 1929 through

[42] Sobel, *The Big Board*, p. 242.
[43] William K. Klingaman, *The Year of the Great Crash – 1929* (New York: Harper and Row, 1989), pp. 175–6. Also see Thomas P. Hughes, *Networks of Power: Electrification in Western Society, 1880–1930* for an analysis of energy systems building.
[44] 84-1 FTC Report 158.

Table 2.3. *Total purchases of Mo–Kan common stock, 1929–1930*

	Mo–Kan Shares			New York Curb Exchange			Chicago Stock Exchange		
Month	Traded Chi & NY	Purchased by Parish & Co.	% by Parish	High	Low	Close	High	Low	Close
1929									
August	45,400	24,900	54.85	34 ¾	29 ⅝	30 ⅛	35 ⅞	30	30
September	131,900	47,900	36.52	35 ⅜	27 ¾	32 ⅝	34 ¾	29	32 ¾
October	168,515	64,525	38.29	34 ½	8	15 ¾	34 ¼	10	16
November	54,775	32,097	58.60	24 ⅝	13 ⅝	16 ¾	23 ½	13 ¼	16 ¾
December	45,930	24,190	52.67	19 ⅞	16 ¼	18 ⅞	20	16 ½	18 ¾
1930									
January	122,545	27,873	22.70	22 ⅝	18 ⅝	21 ¼	22 ⅝	18 ½	21 ⅜
February	134,035	95,959	71.59	23 ¼	21	23 ⅛	23 ⅜	21	23 ¼
March	146,170	42,916	29.36	28 ⅛	22 ⅜	27 ⅞	28 ⅛	22 ⅝	28 ⅛
April	151,530	83,613	55.18	30 ⅜	27 ⅛	30 ⅜	30 ⅜	27 ½	30 ½
May	492,450	263,267	53.45	35 ¾	29 ⅜	35 ⅝	35 ⅞	29	35 ⅝
June	664,725	472,695	71.11	36 ½	15	23 ⅜	36 ½	21	23 ¼
July	143,890	39,426	27.40	26 ½	19 ¼	21 ½	26 ¼	18 ¼	21 ⅝
	2,301,865	1,219,361	53.00						

August 21, 1930, the brokerage firm traded about 2.3 million shares of Mo–Kan on the New York Curb Exchange and the Chicago Stock Exchange; to support the price, Frank P. Parish & Co. repurchased 1.2 million of those shares.[45]

Parish relentlessly pushed the stock. He offered a quarterly dividend "at the rate of 10 percent per annum" to be paid in stock. He also developed a public relations campaign featuring as its centerpiece a monthly newsletter, the *Mo–Kan Bulletin*. The premier issue appeared in August 1929, and the publication soon reached a distribution level of 100,000 copies per month. Meanwhile Parish publicized the company's pro forma balance sheet. This was not an actual accounting; it reflected the company's estimated financial structure after the sale of the proposed securities. An Arthur Anderson & Company audit of March 31, 1930, revealed total Mo–Kan assets of $11,131,000, plus cash and working funds of $1.7 million. In late May, Mo–Kan issued another pro forma balance sheet (dated March 31, 1930) which showed substantially higher numbers: assets of $68 million, including cash and working funds of $58 million, despite the fact that Mo–Kan never had more than $3.7 million in cash or assets valued at more than $10.5 million. These figures reflected the planned stock issues and repurchases by the Frank P. Parish & Company. The pro forma statement made Mo–Kan stock look like a very attractive buy.[46]

H. L. Doherty, head of Cities Service, believed that Parish was simply "carrying on a stock selling proposition . . . [and] did not know anything about the gas business."[47] But Doherty's observations notwithstanding, Mo–Kan's gas business was increasing. In a little more than a year, its staff had grown from 12 to more than 200 employees located in downtown Kansas City. Mo–Kan's lease area increased from 30,000 to 200,000 acres during the same time. It had inaugurated gas service to Owensboro, Kentucky (July 16, 1930), and transferred its executive offices from Kansas City to Chicago.[48] Although operational headquarters remained in Kansas City, the new Chicago executive offices seemed to fulfill Parish's dream of establishing himself as a member of the utility elite. The new offices also brought

[45] 84-A FTC Report 434–9 and 84-1 FTC Report 158.
[46] 84-A FTC Report 436.
[47] 83 FTC Report 542.
[48] *Mo–Kan Bulletin*, vol. 1, no. 1 (August 1929) pp. 1–2, PECA.

Mo-Kan Bulletin

Published Every Month by the Missouri-Kansas Pipe Line Company

| VOL. 1 | Fairfax Building
Kansas City, Mo. | JUNE, 1930 | 640 Otis Building
Chicago, Ill. | NO. 11 |

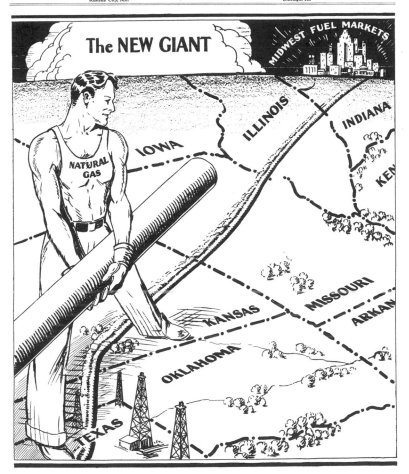

Figure 2.2 *Mo–Kan Bulletin*: "The New Giant," June 1930

him to Chicago's LaSalle Street financial district, where he
would need to be if he was to achieve the next step in his plan
for Mo–Kan.

Panhandle Eastern Pipe Line Company

Despite the collapse of the stock market, neither Parish nor his
fellow executives were inclined to abandon their effort to best
the utility cartel. Their success in marketing shale gas, developing a gathering system, and selling stock encouraged Parish
and his partners to consider an even larger undertaking. The
recently discovered gas fields in the Texas Panhandle and
southwestern Kansas represented a huge untapped fuel supply
which presented unusual opportunities for Mo–Kan's executives. Here they could acquire enormous gas reserves, and if
they could build a pipeline they should be able to tap into major
markets in highly-populated areas in the midwest.

Others saw the same opportunities and they made the Panhandle region a busy place in March 1930. A consortium including Samuel Insull was buying up gas leases for its proposed
pipeline from the Texas Panhandle to Chicago. Mo–Kan was
also extremely active in purchasing leases, particularly during
the first three months of 1930. Much of this acreage Parish acquired through a process known as "checkerboarding," which
entailed a "hit-and-miss" pattern of acquiring leases. Leasing
agents, working under third-party names, scoured gas-producing regions such as the Hugoton Field and bought leases
wherever they could find them. Especially desirable were those
leases adjacent to a competitor's holdings. Mo–Kan acquired
many of its leases under the name Annie L. Millinchip, who was
Parish's aunt. In mid-February, a Mo–Kan subsidiary, the
Texas–Interstate Pipe Line Company, purchased 4,000 acres of
oil and gas leases for $100,000.[49]

By late 1930 Mo–Kan had acquired an estimated 1.5 trillion
cubic feet (Tcf) of natural gas reserves in 150,000 acres of land in
the Hugoton field. Reportedly, the firm controlled leases for
500,000 gas-producing acres in seven states. The reserves in the
Panhandle-Hugoton fields and those in Kentucky were well located with respect to potential markets. As Mo–Kan's public relations director noted, a line drawn from the Panhandle of Texas

---

[49] Panhandle Eastern Pipe Line Company, *Minutes*, February 18, 1930.

to any northern city "would pass through, or reach, through short branch lines, scores of cities and towns where [the] cost of manufactured gas for industrial use is prohibitive."[50]

One of those appealing markets was in the Minneapolis/St. Paul area, where Francis du Pont was trying to negotiate gas sales contracts. Du Pont's efforts failed, in large part because of the competition of the Insull interests, but Parish remained committed to the Minnesota market. At his behest, Brokaw, Dixon, Garner & McKee began mapping a pipeline extending from Liberal, Kansas, to Minneapolis. In anticipation of this project, the Mo–Kan directors had incorporated the Interstate Pipe Line Company on December 23, 1929.[51] Parish could not, however, finance the construction of this 1,000 mile pipeline solely from sales of common stock. Fortunately, he was able to arrange a $20 million line of credit from Continental Bank in Chicago, based on the company's reserves and its apparently healthy balance sheet. While meeting with his banker, Parish discussed his firm's difficulties in negotiating gas purchase contracts with utilities and other industries in the Twin Cities area. The banker told Parish he should meet a particular wealthy, self-described "broker," based in Chicago.

William G. Maguire joins Mo–Kan

William G. Maguire, a successful businessman who knew many of the utility industry's leaders, would be able to provide

---

[50] "A Vast Potential Market for Natural Gas in the Mid-West," *Mo–Kan Bulletin*, vol. 1, no. 10 (May 1939), pp. 2–4, PECA. Also see Eckles, Pt. I, C. II, p. 11, PECA.

[51] The Interstate Pipe Line Company, *Minutes*, December 23, 1929 and Panhandle Eastern Pipe Line Company, *Minutes*, January 2, 1930. The firm's three directors, S. J. Maddin, Ralph G. Crandall, and Francis G. Rearick, authorized a stock issue of 25,000 shares, 15,000 of which would be preferred stock valued at $100 per share with the remaining 10,000 to be common at par value. The capitalization for the pipeline came from intracorporate financing. Parish, president of both Interstate Pipe Line and Kentucky Natural Gas Company, had Kentucky Natural purchase 14,250 preferred shares of Interstate Pipe Line Company stock for $1.425 million. Although several transactions associated with the incorporation of the Interstate Pipe Line Company appear in the Panhandle Eastern minutes of January 2, 1930, future FTC hearings disclosed that the actual deals were made during the summer of 1930 and backdated to January 2, 1930.

Mo–Kan with some of the contacts, and a great deal of the business acumen, Parish lacked. Born in Franklin Grove, Illinois, in 1886, Maguire had attended Lewis Institute. Making a brief visit to St. Louis to attend the World's Fair, he decided to stay and go into business there. The city on the Missouri and Mississippi Rivers would become his home for the next twenty-five years. He first went to work in the offices of the Burlington Railroad. As a young freight clerk, he impressed his supervisors and coworkers with his remarkable recall; he memorized the freight tables of all the railroads in the Midwest. While most clerks consulted tables to determine rates for a car of coal or iron, Maguire could recite the figures from memory.

He left the railroad business for a job with a coal coking firm, one of the businesses that made extensive use of the railroads. In 1918, he became executive vice-president of American Coke and Chemical Company. By 1923, he was president, and he retained this title after the firm changed its name to the St. Louis Gas & Coke Corporation in 1927. Maguire was instrumental in applying the "Roberts method" of coking previously "uncokeable" coal in the years when St. Louis was becoming a major American steel manufacturing center.[52]

At St. Louis Gas & Coke, Maguire met several men who were becoming significant factors in the utility business. He knew Clement Studebaker, Jr., a Chicago broker and son of the founder of the automobile company of the same name, as well as Louis E. Fischer. Studebaker owned a substantial interest in North American Light & Power Company and Fischer was the firm's vice-president. Maguire subsequently purchased an interest in the utility firm, and both Studebaker and Fischer would later provide assistance to the Missouri–Kansas Pipe Line Company.[53]

Maguire left St. Louis Gas & Coke in 1928 at the age of forty-two. He returned to Chicago's financial district, where he opened William G. Maguire and Company, a brokerage firm specializing in the acquisition and development of midwestern utility companies. One of the firm's most lucrative transactions was the sale of natural gas properties to Philip Gossler's Columbia Oil. Maguire's brokerage commission on the sale was sub-

[52] Previously, Maguire was associated with Arthur Roberts, the inventor of the Roberts oven, who was also developing by-products from coal and coke ovens.
[53] 82 FTC Report 302 and 83 FTC Report 290.

stantial enough that he contemplated another retirement in 1930, having several million dollars in cash and an interest in working on Republican political campaigns.[54]

After meeting Parish in 1930, however, Maguire changed his plans. The change did not stem from personal friendship. Maguire's business acumen impressed Parish, but Maguire perceived his future partner to be "a bumptious, unpleasant type of person."[55] Still, Maguire listened. He heard Parish's "intriguing proposition" about a proposed pipeline which would carry gas from the Texas Panhandle to the Twin Cities. Parish described his substantial gas reserves in Texas, Kansas, and Kentucky and asked Maguire to meet with Mo–Kan's engineer, Faison Dixon. Maguire agreed and went to Parish's office, where he spent the rest of the day discussing plans for the pipeline. Dixon told Maguire that Mo–Kan had enough cash on hand, about $17 million, but needed gas sales contracts before beginning to build the line. The situation attracted Maguire, who realized that he could probably use his ties to midwestern utility officials to obtain the necessary contracts. Maguire agreed to help Parish, and he began purchasing Mo–Kan stock for himself.[56]

At first glance, the two men made an unlikely pair. Maguire was quite short and rather blunt in manner, although his insight into business matters brought him respect. Parish, handsome and youthful, had an ambition which outran his expertise. But in several ways, the men had much in common. Both had grown up in the Chicago area, were natural salesmen, and had become wealthy at relatively early ages. Parish made a small fortune but lost it before making another; Maguire's holdings grew steadily yet rapidly to millionaire status. Neither man was satisfied with the wealth or prestige he had acquired – and that provided significant common ground for a deal.

Soon after the meetings with Parish and Dixon, Maguire met with his former associate, Clement Studebaker, Jr., to determine the merits of both Parish and his plan for a pipeline.

---

[54] William G. Maguire, interview by Robert B. Eckles, April 5, 1956, PECA. Also see "William G. Maguire – Natural Gas Pioneer," *The Trunkliner* (Winter 1965), PECA; and "W. G. Maguire Resigns as Gas Company Head," *Gas Age–Record*, February 4, 1928, p. 157.

[55] William G. Maguire, interview by Robert B. Eckles, April 4, 1956, PECA.

[56] 82 FTC Report 298–9.

Studebaker had heard of Parish and his efforts to build a pipeline, and he told Maguire that Odie R. Seagraves and William L. Moody III (Moody–Seagraves Interests) had planned a similar line from Seagraves's Hugoton gas field properties to Omaha. The North American Light & Power Company, the Lone Star Gas Company, and United Light & Power Company had joined forces to purchase the Moody–Seagraves project, rename it the Northern Natural Pipe Line Company, and extend it to Minneapolis. With that in mind, Studebaker suggested that Maguire convince Parish to change the route of the Mo–Kan line.

Maguire also consulted with Louis E. Fischer, vice-president of North American and soon to be named president of the proposed Northern Natural pipeline. Fischer, backed by the likes of Insull, also strongly suggested that Parish shift the route of the Mo–Kan line toward Indianapolis. In return, Fischer told Maguire, North American's subsidiaries in Missouri and Illinois would purchase Mo–Kan's gas.

Maguire presented this offer to Parish. Powerful utility interests had corralled the Twin Cities market, and they would help Parish by purchasing his gas if he would reroute his pipeline toward the east. Parish recalled, "we were glad to do [it]; and we had the firm of Brokaw, Dixon, Garner & McKee survey the markets in Illinois, Missouri, and in Indiana, up to Indianapolis."[57] Like most such entrepreneurial firms, Mo–Kan was flexible, looking always for its big opportunity. Parish now began to look forward to extending the line toward Detroit. While considering the new route and anticipating vigorous competition, during a car ride to a boxing match, Maguire suggested that the Interstate Pipe Line Company change its name to Panhandle Eastern Pipe Line Company to reflect the positioning of its supply and eastward route. The company's name was officially changed on April 28, 1930.[58]

Maguire and Fischer quickly negotiated written contracts (May 24) with North American Light & Power Company. Panhandle would supply natural gas to the latter's facilities in the Illinois municipalities of Springfield, Decatur, Urbana, and Danville. In Missouri, Panhandle would supply gas to North American Light & Power facilities in the towns of Mexico and Warrensburg. Maguire also signed contracts with the Illinois

[57] Ibid., p. 180.
[58] Panhandle Eastern Pipe Line Company, *Minutes*, April 28, 1930.

Power & Light Corporation and the Missouri Power & Light Company. Parish and Maguire appeared to be on the verge of success, despite the downturn in the economy.

But then Mo–Kan's entrepreneurs began to hit some serious resistance to their gas sales pitch. Contracting for gas sales in Indianapolis proved to be difficult because the Citizens Gas Company of that city was operating as a manufactured gas utility and the utility's owners were not willing to write off their investment in coal and coal gas plants. Maguire tried to buy the utility with the idea of converting it to natural gas. But his offer was turned down and Panhandle Eastern lost what could have been a substantial customer.[59]

Tension mounted as Panhandle pressed forward, disturbing the cartel's arrangements. When Maguire had that fateful meeting with Philip Gossler of Columbia Gas & Electric, he learned how difficult the cartel could make his company's affairs. In an attempt to find a compromise, Maguire suggested the possibility of connecting the proposed Panhandle Eastern line with Columbia's gas pipeline network. But Gossler told him that Columbia would oppose any new line into Indiana that it did not own. The cartel, Gossler said, opposed the Panhandle Eastern line and considered it a "raid" against its markets – including those which the large utility firms were not yet serving. The cartel would fight, he warned.

Maguire did not worry excessively about Gossler's warning because he believed he had the support of Fischer and Studebaker.[60] Besides, he had the Illinois and Missouri gas sales contracts in hand and they provided Mo–Kan with the guarantee it needed to move forward with the project. The day after the North American Light & Power contract was signed, Mo–Kan ordered pipe and supplies and contracted for construction of the pipeline and compressor stations. The firm ordered pipe from National Supply Company – as Mo–Kan had earlier – for a price of $25 million. The contract provided for payment of one-half of the purchase price by the tenth of the following month, with terms arranged for the remaining half.

---

[59] Panhandle Eastern Pipe Line Company, *Minutes*, May 23, 1930; August 8, 1930. Also see Thomas A. Rumer, *Citizens Gas & Coke Utility: A History, 1851–1980* (Indianapolis: Citizens Gas & Coke Utility, 1983), pp. 113–18, 155–8.

[60] 82 FTC Report 298–300.

When the proposed long-distance pipeline was announced to investors and the public, the news media reacted with optimism. The downturn in the economy was generating widespread concern, and news of an investment of this size was encouraging. *Time* magazine noted Mo–Kan's endeavors with an article in its issue of June 9, 1930: "Last week one young industry prospered and expanded. . . . For when Missouri–Kansas Pipe Line Company announced that it would build a 1,250-mile natural gas pipe line from the Texas Panhandle to Indianapolis, it evidenced not only increasing activity in the natural gas field but also helped to stabilize a sagging steel production through huge orders for steel. . . . The new M–K line will outdistance the 700-mile Insull–Doherty Amarillo line from Texas to Chicago."[61] Two days later, a group of Texas businessmen feted Parish at a testimonial banquet in Dallas. Bankers, attorneys, and financiers paid homage to Parish as "one who is vitally interested in the progress of this State, and has a growing confidence in its future development."[62]

Within a week, however, the optimism began to fade as the next chapter in Panhandle and Mo–Kan's struggle with the cartel started to unfold. Panhandle Eastern's stock price began to fall. Through the mid-summer of 1930, Frank P. Parish & Company's strategy of "pegging" Mo–Kan stock through stock repurchases had succeeded in bolstering the trading price. But there had been at least one "bear" raid on the firm's securities by unknown groups who sold short a large quantity of Mo–Kan stock. Parish's brokerage firm purchased all of these shares, stabilizing the market at that time. But this was only the beginning of what threatened to be a fatal struggle.[63]

Early in June, a massive and sustained sell-off of Mo–Kan stock occurred on both the New York and Chicago exchanges (Table 2.4). Frank P. Parish & Company again attempted to purchase all of these shares but quickly found itself "taking down" more stock than it could afford to buy.[64] In the meantime,

---

[61] *Time*, June 9, 1930, pp. 43–4. Also see, *Time*, March 17, 1930, p. 50.

[62] "Texas Honors Mo–Kan President," *Mo–Kan Bulletin* (July 1930), p. 6, PECA.

[63] Frank Parish, interview by Robert B. Eckles, July 7, 1957, PECA. Frank Parish later defended his actions when he said stock repurchases were simply the way business was done: "Including the United States Treasury, supporting the bond market – its own bond market – it was essential."

[64] 84-A FTC Report 439.

Table 2.4. *Mo–Kan stock bear raid, June 1930*

Date	Parish & Co. purchases of Mo–Kan stock in excess of its sales
June 6, 1930	8,789
June 7	42,440
June 9	35,842
June 10	55,420
June 11	17,607
June 12	43,459
June 13	24,103
June 14	9,125
June 16	94,677

Source: 84-A FTC Report 439.

William Maguire had met several times with Louis Fischer, who warned Maguire and Parish that the cartel – the "powers that be" – would not permit construction even of the rerouted Panhandle Eastern line. On June 14, 1930 in Chicago, Fischer met once again with Maguire and Parish. Parish suggested that the three spend the afternoon meeting on board the *Theo*, his 150-foot yacht.[65]

Fischer, who had just arrived from New York City, brought Parish another harsh warning. Henry Doherty, Christy Payne, and Samuel Insull were determined to stop the Mo–Kan pipeline project, he said. The three men represented a majority interest in the Natural Gas Pipeline Company's proposed line from the Hugoton field to Chicago, and they considered Panhandle Eastern's line through Illinois as unwelcome competition. Doherty and Payne were already angry at Fischer for assisting Maguire's efforts to sign gas sales contracts for Mo–Kan in Illinois and Missouri. Fischer warned Maguire that pressure would be brought on the Illinois Public Service Commission to oppose Mo–Kan's application for a certificate of public convenience and necessity.[66]

Parish listened carefully to Fischer, who proposed a new

[65] 82 FTC Report 313.
[66] Ibid., pp. 195–6.

plan. Fischer suggested that Mo–Kan surrender gracefully. All that Parish had to do was sell Mo–Kan's gas reserves in Texas to Standard Oil of New Jersey and all of its pipeline properties to Cities Service Company. In return, Fischer said, Mo–Kan stockholders would receive as much as a 35 percent interest in Northern Natural and 10 percent of Natural Gas Pipe Line.[67]

Parish, however, was not prepared to yield. Not yet, at least. He refused even to consider the offer. Then Fischer played his next card. He understood well the powerful forces aligned against Mo–Kan. Unbeknownst to Parish, other utility firms had been working along with the cartel to depress Mo–Kan's stock price. Fischer offered Parish a final warning – a massive bear raid on Panhandle stock would occur on Monday, June 16, unless Parish agreed to stop the pipeline project. Parish countered with an offer to split the cartel. Unwilling to give up his valuable investment in Panhandle's vast gas reserves, right-of-way, and the partially completed pipeline, Parish suggested an early morning meeting on June 16 with Christy Payne of Standard Oil. Parish hoped that by meeting before the stock market began trading, he could convince Payne to call a halt to the raid. Fischer agreed that there might be merit in such a meeting.

Parish and Maguire left Chicago on June 15 on board the Century train to New York City. But their plan went awry when the train was late. The two men did not arrive at the Vanderbilt Hotel until after the stock market had opened on the morning of the sixteenth. Maguire repeatedly called Payne's office, but his calls were not returned until four o'clock in the afternoon, after the market had closed. During the day, Mo–Kan stock tumbled from a high of 36 ⅛ to a low of 15.

Maguire and Parish nevertheless went to the Standard Oil offices late in the afternoon to see Christy Payne. Payne, the fifty-six-year-old president of Peoples Natural Gas Company (a Standard Oil [N.J.] subsidiary) was in charge of all of Standard Oil's natural gas operations. This was the same position his father, Calvin Payne, had held earlier in the century. Payne – "gentlemanly authoritative, polite, well-organized, and meticulously dressed" – greeted Parish and Maguire, along with his business associate, Reed Carr. Payne said: "Well, Parish, how did you like it?" There was no need to explain that he was talking about the devastating decline in Mo–Kan stock. Payne

[67] Ibid., p. 310.

asked if Parish would continue the Panhandle Eastern project –
about 200 miles of pipe had been constructed – after such a serious financial setback. But Parish and Maguire had heard
enough, and they excused themselves and left without comment-
ing on their plans.[68]

As both men knew, the financial setback that day had been
devastating. Early in the day, Parish & Company had purchased the Mo–Kan stock being dumped on the market, but
this time the broker was overwhelmed by the short-sellers. By
the time Parish & Company was holding 50,000 shares in
excess of its own sales, Parish attempted to stop the buying
program. Before he could reach all of his agents by telephone,
however, Parish & Company was holding 94,677 shares of
Mo–Kan stock, which had closed at 21 ½, sixteen points lower
than it began the day. The stock had hit fifteen hours earlier
but rebounded due to Parish & Company's purchases.[69]

As of the close of trading, Parish & Company's liabilities for
Mo–Kan stock totalled $28,123,276.28. The brokerage firm's
underwriting agreement with Mo–Kan stipulated that it pay
cash for stock purchased, but it could now pay for only
$8,045,612.79 worth of stock. There were as well other liabili-
ties arising out of Parish's mishandling of the stock. The end
result was that Mo–Kan was short $19,299,000.[70] The next
day, a Chicago newspaper eulogized Mo–Kan:

> In the case of the Missouri–Kansas Pipeline, it appeared that
> the price was an artificial one maintained for the express pur-
> pose of aiding distribution of the stock. . . . In a general advanc-
> ing market, this might not have appeared for a long time, but in
> the bad markets which we have been experiencing recently, the
> time arrived sooner than some expected. According to reports on
> LaSalle Street, the sponsors attempted to postpone the inevita-
> ble day by delaying delivery of stock, but last week pressure was
> brought to bear and finally, late in the week, delivery was begun

[68] 82 FTC Report 315 and 84-1 FTC Report 161 and "Vote Wide Inquiry on
Short Selling," *The New York Times*, March 5, 1932. Also see Hax Mc-
Cullough and Mary Brignano, *The Vision and Will to Succeed: A Centennial
History of the Peoples Natural Gas Company* (The Peoples Natural Gas Com-
pany), p. 67. Christy Payne also served as a director of Columbia Gas &
Electric Corporation from 1931 through 1933.

[69] 84-A FTC Report 439.

[70] Ibid., p. 440.

of the stock. It was then only a matter of hours before sufficient stock came upon the market to make the burden of support too great for anyone to carry.[71]

Parish, Maguire, Mo–Kan, and the Panhandle venture were all close to defeat, too far in debt to think about anything but getting the best terms they could out of their surrender. They had mounted a serious threat to the cartel. But they had, it seemed certain, been soundly defeated. Their experience in these years is instructive of what entrepreneurial businesses must do as they compete against established monopolistic firms in a relatively unregulated setting.[72]

Although the ultimate success or failure of a capitalist economy depends on the abilities of business people to read and respond creatively to market signals, those signals involve more than prices. The signals emitted by the large utility firms said that they dominated their markets and would do everything in their power to crush Mo–Kan's competition. The power exercised did not involve price competition based on economies of scale and scope. The cartel wanted to forestall entry to prevent competition of that sort, and it appeared that the combine had been successful.[73]

---

[71] "Curb Prices Drop Under Powerful Bear Selling," *Chicago Herald and Examiner*, June 17, 1930.

[72] See Chandler, *Scale and Scope*, p. 34. Chandler discussed the "first mover" advantages of oligopolistic firms in competing with late industry entrants. Chandler's model generally describes the scenario of Panhandle Eastern's unsuccessful attempt to enter the unregulated industry, but Panhandle's history indicates very clearly that Chandler understates the extent to which oligopolistic firms use their political as well as economic powers to block entry.

[73] There are two major opposing viewpoints on the creation of economic markets. Alfred D. Chandler, Jr.'s work, principally *Strategy and Structure: Chapters in the History of the American Industrial Enterprise* (New York: Anchor Books, 1966) and *Scale and Scope: The Dynamics of Industrial Capitalism* (Cambridge, Mass.: The Belknap Press of Harvard University Press, 1990) discusses the primacy of managerial decisions in creating markets: "I focus on the decisions within the institution that led to changes in production and distribution, rather than on changes in the broader economy as indicated by economic statistics – changes that resulted from such decisions" (*Scale and Scope*, p. 9). F. M. Scherer's discussion of industrial organization analysis provides a broader and more convincing analysis of market competition: "In the field of industrial organization, we try to ascertain how market

Parish, Maguire, and their associates had recognized the opportunities to purchase gas reserves and build a pipeline to sell those reserves to cities and industrial firms anxious for natural gas. The opportunity was clearly there. But the large utility firms were beating back this independent entrepreneurial effort by collectively building competing lines and manipulating the financial markets. Parish was also manipulating the stock market, but he had less power than the cartel. As this narrative suggests, the public concern in the 1930s about concentration of economic power was well founded.

Nevertheless, the pipeline industry is necessarily oligopolistic, if only because pipelines are highly capital intensive and one, but probably no more than three, are necessary to serve the largest metropolitan gas markets. Competition would normally be limited to two or three firms. But in the 1920s, even that sort of strategic competition did not exist. The same firms which owned the pipelines typically owned the local gas distribution systems, and even if an independent line were built to serve a particular distributor, the distributor would reject the supply in favor of gas from an affiliated firm. An independent such as Panhandle Eastern found it impossible to compete in such a situation, and this was particularly true when the large holding companies in several adjoining states worked together to prevent entry.[74]

The 1920s interstate utility industry was controlled by market forces, not regulators, but the monopolistic nature of the industry meant that the competition – strategic or otherwise –

processes direct the activities of producers in meeting consumer demands, how these processes may break down, and how they can be adjusted (e.g., through government intervention) to make actual performance conform more closely to the ideal." *Industrial Market Structure and Economic Performance*, 2d ed. (Boston: Houghton Mifflin Company, 1980), p. 2. Scherer, unlike Chandler, accepts the notion that government regulation can be an integral part of a competitive market.

[74] See Stephen G. Breyer and Paul W. MacAvoy, *Energy Regulation by the Federal Power Commission* (Washington, D.C.: The Brookings Institution, 1974), p. 5: "The interstate pipelines have some of the characteristics of natural oligopolies – economies of scale in transmission would seem to justify no more than two or three pipeline sources of supply in any regional market with populations less than 10 million." Also see Paul W. MacAvoy, *Energy Regulation by the Federal Power Commission* (Washington, D.C.: The Brookings Institution, 1974).

which normally resulted in the lowest possible prices and efficient service was not operative. Unregulated market forces had worked to create a new enterprise – Mo–Kan and the Panhandle project – and to provide the utility oligopoly with just the sort of competition needed to bring about innovation. But the result had been monopolistic practices by an intransigent combine. This indicated to some Americans that market forces alone could not be counted upon to provide an efficient, productive American economy. As a result of experiences like this – and others – the public utility industry, the Power Trust, became the most visible symbol of big business in America during the late 1920s and early 1930s. The Power Trust came to represent some of the worst aspects of the American economy and helped to launch a determined search for political solutions to those problems.

3

The Power Trust at bay

When Parish and Maguire left Christy Payne's office and descended to the lobby of the Standard Oil building, they knew they were on the brink of defeat. In front of the statue of John D. Rockefeller, they briefly discussed their bleak situation and then decided to return to the Vanderbilt Hotel. At the hotel, Maguire found six messages from Philip Gossler and learned that Gossler wanted to meet with them at his office by 6:30 that evening.[1]

Philip Gossler was an old hand in the utility business. He had worked in the industry since 1892, after graduating from Pennsylvania State College. His first employer was the United Electric Light & Power Company in New York City, a business which had been in electric power from its birth. He later worked for the Royal Electric Company of Montreal for several years before becoming a director and vice-president of J. G. White & Company, Inc., an electrical engineering firm. From 1909 on, he was the chairman of the board of Columbia Gas & Electric Company, a firm which provided gas and electric service principally in Ohio, Pennsylvania, and New Jersey.

As Gossler knew, he now had the upstart pipeline entrepreneurs in hand. When Maguire and Parish arrived at his office, he merely noted, "You boys have had a hard day." Parish was

[1] 82 FTC Report 316.

50

overwhelmed by the financial beating he had taken, but he tried to retain his poise. "Not too bad," he said, but all these men knew otherwise.[2] Maguire recounted the events of the day and insisted that the Panhandle Eastern pipeline remained a viable project. But Mo–Kan needed $41 million to complete the line, which was still in the early stage of construction. Gossler proposed a deal in which Columbia would have 75 percent ownership of Panhandle Eastern and Mo–Kan would have 25 percent. Parish worked Gossler down to 50 percent, and four days later, on June 20, 1930, Columbia signed an option to purchase from Mo–Kan a one-half interest in Panhandle Eastern. Then, Columbia's auditors spent an entire month examining Mo–Kan's books.

Columbia was a powerful public utility holding company which had for some years either acquired competitors or driven them out of business. For Gossler, this was just business as usual. Columbia had done the same with the American Fuel & Power Company, which produced gas in Kentucky and planned to deliver it to midwestern markets including Toledo and Detroit. Columbia acquired a controlling interest in American Fuel & Power in 1930, prior to the construction of a pipeline, and promptly abandoned the project. Subsequently, Columbia allowed American Fuel & Power to fall into receivership and default on its bonds. That history did not suggest that the future was secure for either Panhandle Eastern or its leaders, Parish and Maguire.[3]

In fact, the months ahead would in many ways resemble a soap opera version of American business. There would be fist fights and charges of fraud, a sudden flight to avoid prosecutions, and an extended episode of corporate espionage. The Panhandle pipeline – now under Columbia's wing – would destabilize a cartel already tottering as a result of the collapse of the stock market and the downward spiral of the national economy. Business as usual would no longer suffice, but the executives heading the industry's largest firms would be slow to recognize that there could be a major trade-off between economic and political power. By the end of the decade and of the nation's worst depression, they would all understand that

---

[2] See 82 FTC Report 201 for Parish's version of this conversation and see 82 FTC Report 315-16 for Maguire's version.

[3] 84-A FTC Report 563, 597–8.

process much better and would be learning how to compete in a new type of political economy.

In 1930, neither Gossler nor his new colleagues, Parish and Maguire, could foresee how much this industry would change. For Gossler, the immediate problem was to absorb his new acquisitions and to pacify the rest of the cartel. He knew that Cities Service and Standard Oil were extremely dissatisfied with his decision to exercise the purchase option for Panhandle Eastern stock. Gossler delayed, asking Parish and Maguire for repeated extensions on the option, extensions which they had no choice but to accept. Finally, Columbia exercised its purchase option on the Panhandle Eastern line on September 17, 1930. Columbia Oil & Gasoline Corporation, a subsidiary of Columbia Gas & Electric, purchased 50 percent of Mo–Kan's stock interest in Panhandle Eastern for $9.8 million.[4] The agreement stipulated that if one of the two firms defaulted on any Panhandle Eastern notes, the other could purchase the remaining one-half interest. This provision was of concern primarily to Parish, Maguire, and Mo–Kan; there was little chance Columbia Gas would default.[5]

Columbia Gas had formed Columbia Oil & Gasoline Corporation four months earlier in order to separate some of its distribution operations from its nonutility companies, presumably to avoid state regulation. After its creation, Columbia Oil acquired the assets of five oil and gasoline subsidiaries as well as the Panhandle Eastern stock. Since Columbia Oil did not have working capital of its own, Columbia Gas handled the financing and provided its share of the funds needed for the construction of the pipeline.

While the new project threatened to disrupt the cartel, Columbia tried to preserve its alliance with the other major utilities. Under the agreement with Parish and Maguire, Panhandle Eastern was to sell 20 MMcf/d to Mo–Kan for resale. This was an important provision because the profits from this business would represent a substantial portion of Mo–Kan's income. Charles Munroe, a Columbia Gas director, asked Parish his intentions for marketing the 20 MMcf/d, adding: "Well, I do hope you won't disturb the Insull people in Indiana." Parish

---

[4] The closing date for the transaction was October 23, 1930.

[5] Securities and Exchange Commission, "Columbia Gas & Electric Corporation et al.," no. 1642, October 2, 1942, p. 220.

assured Munroe that he would not do anything without consulting Columbia first.[6]

Those assurances notwithstanding, Columbia steadily tightened its control of the venture. Maguire arranged for National City to issue $20 million in pipeline bonds to Panhandle and was surprised when Columbia Gas & Electric purchased the entire issue, secured by a first-lien mortgage on the pipeline. Now Columbia held one-half of Panhandle Eastern's stock and 100 percent of its bonds, a total investment of approximately $30 million.[7] At this point, about 200 miles of the line had been built and financed, leaving Mo–Kan and Columbia each to provide $5.5 million to complete the project.

As it became apparent that Gossler had successfully revitalized the Panhandle Eastern line, Columbia's cartel partners became upset. Henry Doherty, the chief executive of Cities Service, was infuriated. One of his company's subsidiaries had actually helped Parish develop a gas market for Mo–Kan in 1928. Subsequently, the upstart firm had broadened its strategy and invaded cartel turf. Doherty wanted the pipeline squelched, not acquired by Columbia. He now tried to block its financing and to prevent the completion of the line.

Henry Doherty was a successful businessman and dangerous opponent. Born in Columbus, Ohio, in 1870, he had left school for good at the age of twelve to become an office boy at the Columbus Gas Co. During the next twenty years, he became a self-taught gas engineer with an expertise in the manufactured coal gas business. In 1905, he organized Henry L. Doherty & Company, which grew rapidly and eventually acquired the control of 190 utility and petroleum properties. In 1930, Doherty received the Walton Clark medal from the Franklin Institute "in consideration of his outstanding and valuable work in [the] development of the manufactured gas industry."[8] Doherty, who had helped promote the Natural Gas Pipe Line Company to serve the Chicago market, was not an

---

[6] Frank Parish Statement. Worth Rowley Files, PECA.

[7] See, Panhandle Eastern Pipe Line, *Minutes*, September 17, 1930; SEC, "Columbia Gas & Electric Corporation et al.," no. 1642, October 2, 1942, p. 222; and Albert F. Dawson, *Columbia System: A History* (New York: J. J. Little and Ives Company, 1938), p. 114.

[8] Ellis, *On the Oil Lands with Cities Service*, p. 158.

opponent of natural gas development, but he was stridently opposed to interference with his markets by independents.

After reading about Panhandle's new financial arrangement in the *New York Times*, Doherty immediately wrote Charles Mitchell, president of National City. Doherty followed up with a rapid series of telegrams expressing his opposition to the deal and attempting to persuade Mitchell not to go forward with it. Doherty described his "great shock" at learning National City would be financing "the Parish natural gas line." National City knew, he said, that the line was "an assault on other utilities." Columbia, he maintained, had a "previous understanding" with Cities Service regarding the threat created by the Panhandle line, but Gossler had gone forward "on grounds of fear that Parish would raid their territory."[9]

Mitchell's response was comparatively brief and simply stated. National City Company wanted to "work out as constructive a solution as possible of a disturbed situation, rendered so by others before it was brought to us for consideration."[10] Doherty responded: "I am still compelled to emphatically object to the part your company has played in this matter and the way your company has played its part. . . ." The new line, Doherty complained, would greatly interfere with Cities Service markets in Kansas. "I think your young men have made a bad mistake . . . ," he warned. "I have tried to be temperate in this matter. . . . There are endless possibilities for trouble. . . ."[11]

While Gossler had to be concerned about this kind of "trouble," he pushed ahead with the project, consolidating his control over Panhandle Eastern and its holdings. Mo–Kan agreed to vest in its pipeline subsidiary all of the Mo–Kan properties except those located in Kentucky, Southern Indiana, and Tennessee which comprised Kentucky Natural Gas. This ensured that the pipeline would have an adequate supply of natural gas. Next, Gossler revamped the board and management of the pipeline. Sam Maddin and Ralph Crandall resigned from the

[9] 84-A FTC Report 272–3.

[10] C. E. Mitchell to Henry L. Doherty, September 23, 1930, in 84-A FTC Report 273.

[11] Henry L. Doherty to Charles E. Mitchell, September 23, 1930; and C. E. Mitchell to Henry L. Doherty, September 24, 1930, in 84-A FTC Report 273–4.

board and were replaced by William Maguire and Francis I. du Pont. Gossler forced Parish to resign as President and Treasurer, moving Fred W. Crawford into those positions. Parish retained his board membership, but the board was enlarged from three to nine members: Mo–Kan and Columbia each had four representatives. Gossler named George H. Howard, president of the United Corporation, which was the largest single stockholder in Columbia Gas, as the ninth "impartial" director. In addition, a three person voting trust comprised of Parish, Gossler, and Howard (as the "impartial" third member) was established to represent the stock interest of Mo–Kan and Columbia. Despite the participation of Maguire and Parish on the Panhandle Eastern board and voting trust, Gossler would call the shots.[12]

Parish and Maguire were now in a tight corner. They urgently needed to come up with $5.5 million, Mo–Kan's share of the remaining pipeline costs. But Mo–Kan was broke, and if Parish and Maguire could not produce the funds, they faced the prospect of losing the pipeline altogether. Determined to hang on, they turned for help to fifty-year-old J. H. Hillman, Jr., who presided over the Hillman Coal & Coke Company. Hillman's father was a Pennsylvania millionaire whose Pittsburgh industrial empire was said to be second only to that of the Mellons.[13] Hillman had formed the investment firm of Pennsylvania Industries through which he acquired interests in the Peoples–Pittsburgh Trust Company and First National of Pittsburgh. Pennsylvania Industries was also a principal stock holder in the National Supply Company, the firm providing pipe for the construction of Panhandle Eastern.[14] National Supply was the sole distributor for National Tube Company, the pipe manufacturing subsidiary of U.S. Steel. Charles Meyer, treasurer of National Supply and a director of Mo–Kan, was anxious for a reputable financier to back Mo–Kan's project, and with that in mind, he had introduced Parish to Hillman.[15]

Hillman wanted references on Maguire whom he believed

---

[12] Panhandle Eastern Pipe Line, *Minutes*, October 20, 1930, and October 23, 1930.

[13] *Forbes*, September 15, 1969, pp. 42–56.

[14] The original Panhandle Eastern line was built with 20-, 22-, and 24-inch segments.

[15] 83 FTC Report 139–43.

was the power behind Mo–Kan. Maguire provided a list of credit references, including most prominently his former associate Clement L. Studebaker, Jr., of Indianapolis. In response to Hillman's inquiry, Studebaker provided a solid endorsement of Maguire's character and acknowledged their close friendship dating back to 1915.[16] This convinced Hillman to support the project, which by this time needed an immediate infusion of capital. Columbia had advanced $500,000, which Mo–Kan had to match. Parish actually persuaded Hillman to advance the same amount prior to the conclusion of their negotiations.[17] Later, after they had sealed their agreement, Hillman, in March 1931, provided the remaining $5 million Mo–Kan needed to finance the construction of Panhandle Eastern.[18]

Gossler was not pleased. The scenario he had in mind was clearly the same one he had developed for the American Fuel & Power Company. But Mo–Kan had refused to roll over and die. Gossler, who was not a man to keep his feelings to himself, wrote Hillman two weeks later: "We can only construe this as an act of aggression on the part of your associates against the best interests of Columbia which, if pursued, we should have to resist to the full extent of our resources and ingenuity."[19] Gossler, now sounding like Henry Doherty, was nevertheless unable to block this new deal. Hillman was a substantial industrialist, and he had the resources to pump life into Mo–Kan, which still owned 50 percent of Panhandle Eastern's stock and remained for the time an independent firm.

[16] Ibid., p. 1019, 148.

[17] Ibid., p. 49.

[18] 84-1 FTC Report 161. The FTC's chief accountant explained the transaction during the FTC public utility hearings: In accord with negotiations with Hillman, the Panhandle Corporation was formed. This new corporation would own 50 percent of Panhandle Eastern's stock. The Corporation issued Hillman "2-year notes with a face value of $4,940,000." The Missouri–Kansas Pipe Line Company issued notes for $1,000,000 and $60,000, for a total of $6 million. They also issued 100,000 shares of class A Missouri–Kansas Pipe Line Company stock; also warrants evidencing the right to purchase 120 shares of class A stock before April 1, 1936, at $15 a share. All of the above mentioned securities were sold [to Hillman] for a consideration of $5,440,000. Dillon, Read & Co. underwrote the issue.

[19] 83 FTC Report 158.

Construction of the Panhandle Eastern pipeline

In that tense setting, Panhandle pushed ahead with the construction of its pipeline. The engineering firm of Brokaw, Dixon, Garner, & McKee had estimated the cost of the line to be $41,261,000, including $25 million for 231,000 tons of pipe. The expected completion date for the pipeline was April 1, 1931. In determining the best route the engineering firm used the most advanced technology to survey the right-of-way. Aerial surveys, telescopic photography, and other innovations provided much greater accuracy of measurement than the traditional preliminary route survey conducted by work crews who walked the entire length of the proposed line. The new technologies substantially accelerated the 900-mile survey to a matter of days rather than weeks.[20]

In the construction, Panhandle Eastern also used innovativemethodstosavetimeandmanpower.Caterpillartractorsrolled through terrain inhospitable to trucks. Ditching machines operated twenty-four hours a day to prepare the way for the pipe-laying crew. Side-boom and backfill tractors lowered pipe into the ditch. Two men operating a backfill machine could cover a mile of ditch for a large diameter pipe in a ten-hour day, doing work which had once been done by hand at a much slower pace.[21] Panhandle Eastern's first two compressor stations were built near Liberal and Louisburg, Kansas. Fluor Corporation, Ltd., of Los Angeles designed and constructed virtually identical stations at each location and planned additional compressors to be built at eighty-mile intervals. Meanwhile, Brokaw–Dixon installed a telephone system along the

[20] Brokaw, Dixon, Garner & McKee, *Report of the Activities of Brokaw, Dixon, Garner, & McKee: February 1, 1930 to April 1, 1931 for Panhandle Eastern Pipe Line Company* (cover letter dated July 31, 1931), PECA. With the gas reserves and marketing surveys successfully completed, Missouri–Kansas Pipe Line Company formally contracted with Brokaw, Dixon on June 24, 1930, to provide general contracting services for the line through the anticipated completion date of April 1, 1931. Brokaw–Dixon hired the Curtis–Wright Flying Service of New York City in May 1930 to photograph four-mile wide strips of ground in the hilly and heavily wooded sections of Missouri and Illinois.

[21] Elmer F. Schmidt, "Modern Pipe Line Construction Methods," *Proceedings of the American Gas Association,* 1929, pp. 168–74.

Figure 3.1 Pipeline survey crew, 1930

Figure 3.2 Horse-drawn pipeline welding equipment, 1930

Figure 3.3 Welders, 1930

line. This private system kept the operational headquarters in Kansas connected with the field and compressor station crews at any time.

Panhandle Eastern faced its first regulatory challenge when it attempted to lay pipe and contract for gas sales to distribution companies in Illinois. In order to receive a certificate of public convenience and necessity from the Illinois Commerce Commission, the firm organized the Panhandle Illinois Pipe Line Company (with a $1 million capitalization), which would, from the perspective of the Illinois commission, be an intrastate pipeline. In this way, the state could extend its regulatory control over a segment of an interstate pipeline, even though this arrangement would not prevent Panhandle from selling gas to its Illinois subsidiary at whatever price it chose. Nine months after the preliminary hearing, the commission issued the certificate on April 8, 1931, and construction began shortly thereafter.[22]

Panhandle Eastern's system was divided up in other ways as well. In Texas, another wholly owned subsidiary of the pipeline firm, the Texas–Interstate Pipe Line Company, operated the gathering system which delivered Texas-produced gas to the Panhandle system. In Indiana, after the line was extended, Columbia Gas sold the portion of Panhandle Eastern's line operating from the Illinois–Indiana border to Zionsville, Indiana to the Indiana Gas Distribution Corporation, a subsidiary of Columbia. Technically, Panhandle Eastern's line then extended from Texas to the Missouri–Illinois border while subsidiaries or affiliated firms operated in Texas, Illinois, and Indiana.

---

[22] Panhandle Eastern, *Minutes*, May 23, 1930; August 8, 1930. For a review of Panhandle Eastern's relationship with the state of Illinois see Illinois Commerce Commission, *Illinois Commerce Commission v. Panhandle Eastern Pipe Line Company*, Docket 38142, December 13, 1951. Also see Earl D. Bragdon, *The Federal Power Commission and the Regulation of Natural Gas: A Study in Administrative and Judicial History* (unpublished Ph.D. Dissertation: Indiana University, 1962), pp. 34–5. Utility captain Samuel Insull actually played an important role in the passage of the 1913 bill, which created the State Public Utility Commission of Illinois. This scenario fits rather well under the capture thesis propounded by Gabriel Kolko in *The Triumph of Conservatism: A Reinterpretation of American History, 1900–1916* (New York: Free Press of Glencoe, 1963), which states that industry promoted regulation in order to stabilize the market dominance of existing firms.

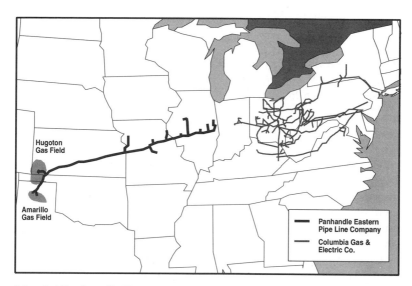

Map 3.1 Panhandle Eastern Pipe Line Company and Columbia Gas & Electric Company, 1931

There were other problems as well. In its efforts to acquire right-of-way along the pipeline route, Panhandle had to negotiate individually with land owners. A former employee, Harry Pope, recalled that "The Farm Bureau, an organization composed of farmers, not only of Illinois, but all over the country ... advised the farmers not to sign right-of-way agreements."[23] Panhandle Eastern officials had to convince the farmers that once the pipe was buried under their land they could continue to cultivate crops on top of it; they strengthened their case by also paying up to $1.00 per rod for right-of-way in Illinois farm land.

The *New York Times* reported the completion of the line on August 21, 1931: "Pipes link the East with gas in Texas" (Map 3.1).[24] After a brief period of testing, the 930-mile Panhandle Eastern system began delivering gas to customers in Kansas and Illinois. The initial capacity was only 175 MMcf/d, powered by three compressor stations with a total of 14,000 horsepower.

---

[23] Harry Pope, interview by Robert B. Eckles, February 6, 1956, PECA.

[24] "Pipes Link the East with Gas in Texas," *New York Times* (August 21, 1931), p. 24.

During the early 1930s, the firm continued to extend the system slowly, consolidating markets in Kansas, Missouri, and Illinois. Although the system extended into Indiana, the mainline terminated in a cornfield seventy-five feet across the Indiana–Illinois border near Dana, Indiana, a site unlikely to either generate large volume gas sales or to anger the other members of the cartel.[25]

Gas markets dry up in the depression

While a technological success, the pipeline was not making any money. In fact, after posting a respectable income in its first year of operation, the Panhandle Eastern system lost money during the next four years. This was especially bad news for Mo–Kan, which looked to Panhandle Eastern as its major source of revenue. In part the problems were a product of the Great Depression. Economic stagnation halted most industry growth during these years, and many pipelines were operating at less than 50 percent capacity. Only three medium length pipelines were constructed between 1932 and 1936, none of them longer than 300 miles; no other major lines were built until 1943.[26]

In the midwest, Chicago was now receiving natural gas, as were Omaha and the Twin Cities. But Indianapolis and Detroit, Panhandle Eastern's proposed markets, remained locked into manufactured gas, as did other midwestern cities. On the West Coast, Pacific Gas & Electric had built a line from Southern California to San Francisco. But this left large supply areas in Texas, Oklahoma, Kansas, and Louisiana with enormous amounts of the fuel unconnected to markets. Texas oil drillers, concerned only about "black gold," continued to vent millions of cubic feet of unmarketable "waste gas" into the atmosphere.[27]

Although Frank Parish knew that the collapse of the econ-

---

25 For an interesting account of the economic development prospects for the booming natural gas industry in small midwestern towns see "Boonville's Gas Comes From Distant Texas Panhandle," *Boonville Daily News*, December 28, 1931, PECA.

26 American Gas Association, *Gas Rate Fundamentals* (American Gas Association, 1985), pp. 14–15.

27 Sanders, *The Regulation of Natural Gas*, p. 24.

omy was holding down demand, he had ample reason to suspect that Columbia Gas was intentionally preventing Panhandle from making a profit. He blamed Columbia for not purchasing the gas it originally contracted from Panhandle Eastern and for not seeking new gas markets. Parish believed that Columbia's intent was to strangle Mo–Kan and then by default acquire complete control of the pipeline. During a board meeting in October of 1931, Parish demanded a statement enumerating management's efforts to ensure the sale of Panhandle Eastern's gas. Company president Crawford told Parish that any such efforts had to abide by the existing territorial constraints: "Any gas to be taken under such contract by Columbia Gas & Electric Corporation would be for supplying Indianapolis and northern Indiana territory, extending to Detroit and other points in Michigan and connecting with existing pipe lines in Ohio." Contracts with "the Missouri–Kansas Pipe Line Company [Mo–Kan] would be only for the purpose of supplying gas to the southwestern part of Indiana, through Terre Haute, and south to Nashville, Tennessee." Crawford insisted that neither Columbia nor Mo–Kan was entitled to gas except on these terms.[28]

Parish disagreed and kept disagreeing. Although the Board passed a resolution during the summer of 1931 calling for gas sales contracts with both Columbia and Mo–Kan, this appeared to be nothing more than an attempt to appease Parish.[29] In August, Parish again chastised the Board, and this time Crawford offered to resign if Mo–Kan was displeased with him. But that would not solve the problem as long as Columbia was unwilling to change its position. Columbia was unwilling to budge, as Director Munroe made clear. He had "grave doubt as to the advisability" of the contracts Parish and Maguire wanted to sign in order to make Panhandle profitable.[30] It was still business as usual for the cartel.

Cornered, Parish tried to put more pressure on Columbia. He brought a court reporter with him to Panhandle's October 27 board meeting, charging that Columbia had not accurately recorded the minutes. Munroe objected strenuously. There was a quick motion to adjourn and the meeting ended abruptly.

[28] Panhandle Eastern, *Minutes*, May 26, 1931.
[29] Panhandle Eastern, *Minutes*, June 5, 1931.
[30] Panhandle Eastern, *Minutes*, August 13, 1931.

Neither Parish nor Maguire attended the next several meetings of the board, and the slate of directors proposed for the
following year did not include either of their names. Parish was
still a member of the three-person voting trusteeship, but in
effect he had no power. Howard and Gossler, both of Columbia,
voted together on every issue, including the new board of directors. Parish and Maguire, it appeared, had played their last
cards and were about to lose both Mo–Kan and the pipeline.

Espionage

Parish knew that Philip Gossler was threatening to drive him
out of business, but what he did not realize was that Henry
Doherty, head of Cities Service, was also committed to that
same goal. Doherty was not about to stand by and watch Panhandle, with or without Columbia's support, expand its system
into his territory. After Cities Service failed to block the National City loan, the firm began to spy on its opponent in an
effort to gain useful information about Parish and Panhandle
Eastern's operations.

The episode began in August 1930, six weeks after the debili-
tating stock raid on Mo–Kan. At that time, W. J. Hinchey, a
former member of Parish's stock syndicate and then a supervisor
at Frank Parish & Company, hired Elsie V. Walker as a secretary and assistant. Little did Hinchey know that Cities Service
Company had enlisted her to seek the job, learn as much as she
could about Panhandle's operations, and report her findings
back to Cities Service. Walker had worked for Hinchey years
earlier when he was employed by Gas Service Company, but she
was let go after Hinchey quit to work for Parish. Later she had
worked in the securities department of Henry L. Doherty &
Company, but she had lost that job in a general layoff during
July of 1930. Her situation was tense after she learned that her
broker required a $600 payment on her securities account or she
would be forced to sell off her existing holdings.[31]

Unable to find another job, Walker was in despair. While she
was sitting at a desk in a Cities Service office (after unsuccessfully seeking employment), she was approached by Thomas W.
Shannon, an internal investigator at Cities Service, who asked
her why she was upset. She explained what had happened and,

[31] 83 FTC Report 47–64.

later that same evening, Shannon and an assistant prosecuting attorney (Olcott) for the City of Cincinnati met with Walker in her room. The two men told her to secure a position with her former supervisor, Mr. Hinchey, and the attorney implied that she would be investigating wrongdoing at Mo–Kan, not spying on a competitor. Shannon agreed to pay Walker $200 a month and help her pay off her stock debt. He also gave her a $100 advance.[32]

Elsie Walker began spying on Frank Parish & Company in August 1930. Although she was let go during a company-wide layoff at the end of September, Mo–Kan rehired her in December 1930. She remained at that job through May 1932, earning $110 a month as a secretary and almost twice as much as a spy. She steadily relayed information to Shannon. When Columbia Gas agreed to invest in Mo–Kan stock, Shannon knew about the deal the evening after the agreement was signed. Whenever Walker's boss directed her to make carbon copies of a document, she used two pieces of carbon paper and passed the second copy to Shannon, as per his instructions. They developed a code for other messages: Parish was "persimmon," Mo–Kan was "lemon," and Mr. Hinchey was "peach."

Shannon was well-informed about Parish and Panhandle. He knew about the first contract between Columbia and the pipelines, the salaries of Panhandle's executives, the activities of its top managers, and on a weekly basis the progress being made on construction of the new line. One of the last items Walker reported to Cities Service was Panhandle Eastern's early attempt to acquire contracts to provide natural gas to the cities of Indianapolis and Detroit. For two years Walker continued to keep Shannon up-to-date on her unsuspecting employers at Mo–Kan.

Early in 1932, however, as Mo–Kan headed toward bankruptcy, Shannon abruptly terminated the espionage. As he explained to his spy, Parish shortly was headed for the penitentiary. In May, 1932, Shannon asked Walker to sign a form in exchange for a final payment of $500 releasing Cities Service Company, its subsidiaries, Henry L. Doherty, Henry L. Doherty & Company, and Gas Service Company, from all obligations and claims to her.[33]

[32] Ibid., p. 50. Most of her subsequent payments went directly to her brokerage and bank accounts.

[33] Ibid., pp. 63–4. Also see "Utilities Hearing Told of 'Spy' Role," *The New York Times*, November 16, 1935. After Miss Walker testified, a Cities Service

Mo–Kan falls into receivership

Doherty was anxious to get clear of Elsie Walker because his firm's campaign to disable Panhandle and ruin Frank Parish was on the verge of success. Chicago newspapers reported in late January and early February that Parish would soon be indicted for mail fraud. On February 13, 1932, Parish, still President of Mo–Kan, launched a preemptive strike and filed an antitrust suit against Cities Service and Standard Oil for $75 million claiming that those organizations had attempted to destroy Mo–Kan in 1930. By this time, however, Mo–Kan was about to slip into insolvency leaving Panhandle in Columbia's hands.[34] On March 6, 1932, he petitioned for the receivership of Mo–Kan, a maneuver which would at least forestall any attempt by Columbia to gain complete control over his pipeline firm. On March 18, 1932, Delaware Court chancellor J. O. Wolcott deemed that Mo–Kan, along with its only operating subsidiary Kentucky Natural Gas, had defaulted on their bank notes and were insolvent. Wolcott appointed two receivers, O. Ray Phillips of Chicago and Henry T. Bush of Wilmington, to direct Mo–Kan's operations.[35]

Maguire also fought back and urged the Mo–Kan receivers to file an antitrust suit against Columbia for deliberately leaving Panhandle in poor financial condition and thus forcing Mo–Kan into bankruptcy. According to Parish, the pipeline had 40 MMcf/d of excess capacity and needed to sell at least an additional 15 to 20 MMcf/d for about $.20 per Mcf to remain viable.[36] Both he and Maguire were certain it could do so if Colum-

official implied that the espionage was only for identifying those high level employees that Parish was attempting to hire for his own organization and then try to keep them on the Cities Service payroll.

[34] Eckles, C. III, p. 68, IHS. Subsequently, Mo–Kan sued Columbia for $180,000,000; this suit was later settled for $200,000; Mo–Kan, *Minutes*, March 18, 1942. See "Hints More Utility Suits," *The New York Times*, February 17, 1932; "U.S. Eyes Mo–Kan," *Chicago Evening American*, January 18, 1932; and "Columbia Gas Sued for $180,000,000," *The New York Times*, July 19, 1935.

[35] "Appeal of Missouri–Kansas Pipe Line Company from a decree entered on the 9th day of October, A.D. 1940, by the Chancellor of the State of Delaware in *Missouri–Kansas Pipe Line Company v. Dupuy G. Warrick*, PECA." Also see *The New York Times*, March 2, 1932.

[36] Brief and argument of certain objectors to the proposed settlement in the Court of Chancery of the State of Delaware. RE: Missouri–Kansas Pipe Line

bia was not blocking its growth. While Panhandle Eastern's financial situation was not unique, the geographical setting of its pipeline with the possibilities for extension should have offered it ample opportunities to make money.[37]

While these struggles were taking place, J. H. Hillman commenced foreclosure proceedings on Kentucky Natural Gas. Hillman went after Kentucky Natural because it was the only operating company still controlled by Mo–Kan. Hillman's foreclosure action allowed him to reorganize Mo–Kan and become Mo–Kan's new representative on the Panhandle Eastern board. Hillman also successfully negotiated natural gas sales contracts with Indianapolis area manufacturing firms supplied by Kentucky Natural Gas through the Panhandle pipeline.[38]

Now Parish's world rapidly crumbled around him. On March 24, 1932, a federal grand jury in Chicago handed down a forty-nine-count indictment against Parish, Maddin, Ralph Crandall, and J. T. McManmon, all officers and directors of Mo–Kan. The principal charge was mail fraud, and the grand jury charged the defendants with misrepresenting Mo–Kan's financial condition in materials mailed to investors. Parish and his co-defendants were accused of selling through misrepresentation Mo–Kan stock to 20,000 investors for as much as $50 per share; at the time of trial the stock was worth $.50 per share. Another charge involved the company's representation to the public that E. I. du Pont de Nemours Company and its subsidiaries owned a substantial quantity of Panhandle Eastern stock. The judge scheduled the trial to begin in early 1933.[39]

Three weeks prior to the trial, Parish's attorney abruptly resigned and Parish unsuccessfully attempted to have the trial postponed. Failing to obtain a legal postponement, Parish, in

Company, Frank P. Parish & Company, Kentucky Natural Gas Company, and Indiana-Kentucky Natural Gas Corporation, p. 42, PECA. Also see "Pipe Line to Receivers," *The Kansas City Times*, March 19, 1932, PECA.

[37] See Ralph K. Huitt, "Natural Gas Regulation Under the Holding Company Act," *Law and Contemporary Problems*, vol. 19, no. 3 (Summer 1954), p. 461.

[38] Eckles, C. III, pp. 69–71, IHS. Also see Thomas A. Rumer, *Citizens Gas & Coke Utility: A History, 1851–1980*, pp. 113, 118, 155–6. Citizens Gas did not begin purchasing natural gas for its distribution customers until 1950.

[39] "35,000,000 Fraud in Stocks Laid to 4," *The New York Times*, March 25, 1932; and "Frank Parish Indicted," March 24, 1932, (Chicago) paper unknown, Frank Parish file, PECA. Mo–Kan director Francis I. du Pont owned about 3,000 shares, which he finally sold in 1934 when the stock was worth very little.

February, followed the example set several months earlier by Samuel Insull and fled the country. Using a false Canadian passport, he traveled to Germany.[40] In Parish's absence, his legal problems increased. At least two more suits were filed against him including one by William Maguire. Maguire, who had received commissions from the deal with Columbia Gas and the bond issue by National City, had set aside a substantial amount of money which he and Parish planned to use as a "war chest" in their battle to keep Mo–Kan and resist Columbia. Believing that Parish had misused those funds, Maguire sued him after Parish fled the country.[41]

In December 1933, Parish voluntarily returned to the United States after learning that his brother was gravely ill.[42] Angry that Maguire had accused him of stealing the "war chest," Parish and a companion confronted Maguire at the Palmer House Hotel. Maguire politely asked the two men to come into his room and offered them drinks, but Parish and his friend began to pummel Maguire. Parish was subsequently arrested for this incident as well.[43]

By early 1934, Mo–Kan was on the brink of extinction, an easy target for Columbia Gas to hit. Mo–Kan was in receivership; J. H. Hillman had stripped its only operational subsidiary; and Parish was under a federal indictment. Columbia Gas officials offered to pay Mo–Kan $300,000 for its interest in Panhandle Eastern – providing that Mo–Kan would not file, as Maguire now threatened, an antitrust suit against Columbia. The Mo–Kan receivers considered the offer, but the Delaware Court of Chancery rejected it. Instead, in July 1935, the Mo–Kan receivers filed an antitrust suit against Columbia in the U.S. District Court, Southern District of New York.

[40] *Chicago Daily Tribune*, February 20 and 21, 1933, PECA. Insull, who was also indicted on a charge of mail fraud stemming from the collapse of his huge public utility empire, had fled the country in June, 1932.

[41] See Eckles, C. III, p. 73, IHS. This suit was later settled when Maguire accepted a combination of notes and bonds, presumably in the name of Parish, from Mo–Kan and Kentucky Natural valued at $350,000.

[42] *Chicago Daily Tribune*, December 5 and 8, 1933. See Eckles, C. III, pp. 72–3, IHS.

[43] "F. P. Parish Gives Up on Beating Charge," *The New York Times*, September 5, 1934. Also see *Chicago American*, September 5, 1934.

Parish was also in court at this time, in a trial which lasted for six weeks. Horace A. Hagen, special assistant U.S. attorney general, contended that Parish had engaged in "one of the greatest exploitations in American financial records." The company's history was "full of fraud – a 'bubble' which burst with a loss of \$35 million to shareholders."[44] During the testimony, however, Parish actually entertained the jury, spinning stories of his youth. He contended that Cities Service had employed various illegal tactics, including stock manipulation and espionage, to halt progress on the pipeline. Parish's attorneys also charged that Doherty "raided" Mo–Kan stock, forcing the firm to seek an unfavorable alliance with Columbia Gas and resulting eventually in the 1932 receivership. This argument apparently convinced the jury that Mo–Kan and its Panhandle Eastern subsidiary were victims of corrupt competition rather than purveyors of financial manipulation. The jury found Frank Parish not guilty on April 30, 1935.[45]

Although Parish avoided federal prison, he would never again exercise direct control of Panhandle Eastern. During the summer of 1935, the company amended its certificate of incorporation once again and reorganized the voting trust. Parish was removed from the trust, and in the meantime, he had dissipated his financial stake in Mo–Kan. By contrast, Maguire had strengthened his holdings of Mo–Kan stock, and he would attempt to revive that enterprise in a political economy transformed by the New Deal.

The Federal Trade Commission investigation

The problems in the utility industries were attracting a great deal of political attention in the 1930s. Struggles such as those over Panhandle Eastern confirmed the fears of those politicians who had long focused their antitrust sentiments on utili-

---

[44] *New York Times*, April 4, 1935. During the proceedings, U.S. District Court Judge John P. Barnes granted defense attorney motions for not guilty verdicts against Parish's associates Ralph Crandall and James McManmon, although the charges against Parish and Samuel Maddin remained.

[45] "Parish Tells His Meteoric Rise into Big Money," *Chicago Daily Tribune*, April 25, 1935. Also see numerous articles about the trial found in the following newspapers throughout April 1935: *Chicago Herald and Examiner*, *Chicago Daily Tribune*, *Chicago Daily American*, and *Chicago Daily News*.

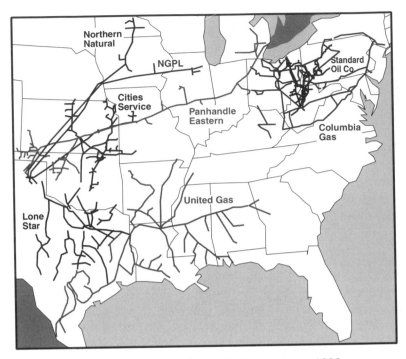

Map 3.2 Major natural gas pipeline systems, 1936

ties (Map 3.2). The economic power and political influence of the "Power Trust" had attracted congressional attention since the early 1920s. Several senators, including George Norris (R–NE) and Thomas Walsh (D–MT), had assisted in launching an intense campaign in favor of government-owned power facilities.[46] These Senators and their supporters prompted the Federal Trade Commission to launch an investigation of the operations of General Electric and its subsidiary, the Electric Bond & Share Company, which in 1924 produced about eleven percent of the electric energy generated in the entire country. While General Electric had divested its interest in Electric Bond & Share by the time the FTC completed its report in 1927, the commission noted with concern the increasing power of Samuel Insull and other members of the cartel. These interests, the

[46] For an account of another public power facility, see Thomas K. McCraw, *TVA and the Power Fight, 1933–1939* (Philadelphia: Lippincott, 1971).

FTC said, dominated the electric and gas businesses within several large regions of the nation.[47]

Senator Walsh pressed for another study, and on February 15, 1928, the Senate directed the FTC to investigate and report on the condition of the existing public utility holding companies.[48] The result was what historian Ellis Hawley has characterized as "one of the most widely publicized and extensive investigations of all time."[49] The FTC – later joined by the House Commerce Committee – produced a massive ninety-six-volume report which was published in 1935. These volumes introduced readers to the "jungles of holding company accounting, [and] some of the weirdest and most disastrous financing the nation had ever known."[50] As the FTC report established, one of the foremost problems was the pyramiding of holding companies, each of which was heavily leveraged with bonded debt. This debt structure left the systems vulnerable to an economic downturn. Pyramiding enabled "a minimum investment to control a maximum of operating facilities, and the result was a variety of abuses."[51] Utility companies also flagrantly inflated their assets to secure a larger rate base and mounted political and public relations campaigns which increased their influence over local government, schools, churches, and the media.[52]

The natural gas industry attracted substantial attention from the investigators. The FTC revealed that four holding companies controlled directly and indirectly more than 60 percent of all natural gas produced in 1934 and 58 percent of the total pipeline mileage in the United States.[53] By 1934, nearly 40 percent of all natural gas crossed state lines, and the fuel

---

[47] See Joint Resolution no. 329, of the 68th Cong., 2d sess., 1925 and the "Electric Power Industry, Control of Power Companies," 69th Cong., 2d sess., issued on February 21, 1927 as Senate Document no. 213.

[48] *Congressional Record*, 70th Cong., 1st sess., 1928, vol. 49, part 3, p. 3054; Congress, Senate, *Senate Resolution 83*, 70th Cong., 1st sess., 1928, as extended by *Senate Joint Resolution 115*, 73rd Cong., 2d sess., 1934. Also see 84-A FTC Report (all); and William E. Leuchtenburg, *The Perils of Prosperity, 1914–32* (Chicago: University of Chicago Press, 1958), pp. 190–1.

[49] Ellis Hawley, *The New Deal and the Problem of Monopoly* (Princeton: Princeton University Press, 1966), p. 326.

[50] Ibid., pp. 326–7.

[51] Ibid., p. 330.

[52] Ibid.

[53] 84-A FTC Report 591 and 611.

was being used in 34 states by 7 million industrial, commercial, and residential customers. These consumers needed, the FTC said, an "adequate, dependable, and safe service at reasonable prices."[54] The FTC report identified sixteen "evils" in the natural gas industry. These included: (1) waste, (2) unregulated pipelines, (3) expensive competitive struggles for markets, and (4) reckless financial manipulation.[55] Out of the report was to emerge a new body of regulatory legislation directed at the various sections of the utility industry.[56]

According to the FTC, the ability of individual states to regulate natural gas was "at best indirect, partial, and poorly founded because of their limited authority to ascertain facts and their lack of authority to regulate interstate commerce."[57] The report outlined the cartel's operations, noting the traditional "system of so-called 'ethics' which holds that it is unfair and unethical for a pipe line to invade any territory already served or claimed by another line. Such invasions are classed by the industry as 'raids' and the fear of retaliation by powerful interests is usually sufficient to preclude them."[58] The report also identified other aspects of the competitive situation Panhandle had encountered, including "a definite policy of many natural-gas companies to regard certain territory as their own, to recognize certain other territory as belonging to others, and to refrain from invading the territory of another company."[59] Given this situation, further regulation was necessary.

The FTC closely examined Panhandle Eastern's brief history of competitive intrigue, finding in the relationship among Cities Service, Columbia Gas, and Panhandle Eastern, a superb example of monopolistic behavior. This case was particularly important, the commission said, because it revealed so clearly how the cartel's "understandings" and "gentlemen's

---

54 Ibid., p. 611.
55 Ibid., pp. 615–16.
56 Also see, Federal Trade Commission, *Investigation of Concentration of Economic Power, no. 36*, published as Temporary National Economic Committee Report no. 76-3 (Washington: GPO, 1940). Hereafter referred to FTC Monograph no. 36.
57 84-A FTC Report 609.
58 Ibid., p. 593.
59 Ibid., p. 276.

agreements" worked in regard to service areas and pipeline routes. The investigators also examined the bear raid of June 16, which forced Panhandle Eastern to seek assistance from Columbia. The Commission's lawyers questioned one suspected leader of the raid, Christy Payne of Standard Oil (N.J.). "I would like to say emphatically that I never traded any stock of the Missouri–Kansas Pipe Line Co.," Payne said.[60] He also suggested that no raid took place, but the Panhandle Eastern case nevertheless came to symbolize the problems which existed in the gas industry.

To prevent such wanton abuse of corporate power, the FTC proposed greater government oversight:

> A federal regulatory law should be enacted applicable to inter-state natural gas pipe lines which transport gas for ultimate sale to and use by the public, regulating contracts for purchase of gas to be transported interstate, or regulating rates for carriage or city gate rates at the end of such transportation, or all of these.[61]

The Commission also made recommendations about the need for conservation and the elimination of gas wastage; a provision for federal authority to order reasonable extensions of service; the need to separate the electric utility and gas line businesses; and a law to restrict banks to investing in, rather than managing, natural gas firms.[62] The tone of the report, the collected evidence, the antibusiness sentiment of the depression years, and the problems of the utility industries made it virtually inevitable that a more formidable regulatory system would be created. This was especially true by the middle of the decade, after the NRA experiment with government sponsored cartels had collapsed.

New Deal legislation and the utility industry

Franklin Roosevelt was far more flexible than his predecessors in allowing the federal government to intervene in the

---

[60] 83 FTC Report 279 and 282–3. Also see "Payne Denies Raid on Pipe Line Stock," *The New York Times*, November 22, 1935.

[61] 84-A FTC Report 616–17.

[62] Ibid.

American economy, particularly in those sectors of the economy which did not seem to be working. With Roosevelt's support, Congress would explore new regulations for a number of industries including banking, securities, airlines, trucking, and utilities.[63]

Roosevelt was a particularly strong proponent of public power and regulation of the utility industry. During the 1932 presidential campaign, he had attacked Samuel Insull and the "evil" power of the electrical power industry. Previously, as governor of New York, Roosevelt had waged a campaign for cheap electric power.[64] In his State of the Union address in January 1935, he called for the abolition of holding companies, and gave Texas congressman Sam Rayburn and Burton K. Wheeler of Montana the task of introducing legislation to achieve that end. Rayburn was an ardent opponent of the utility cartels and their banking habits, and he helped FDR achieve his legislative objectives.[65]

Given the political and economic environment of the 1930s, it was almost certain that Congress would impose new regulations on the natural gas industry. It was not at all clear exactly what form the regulations would take, and it was not clear that

[63] Schwarz, Jordan A. *The New Dealers: Power Politics in the Age of Roosevelt* (New York: Knopf, 1993), p. xi. For a very readable account of the emergence of regulation, and the men who guided several regulatory agencies, see Thomas K. McCraw, *Prophets of Regulation*, (Cambridge: The Belknap Press of Harvard University Press, 1984.) For a review of the origins of trucking regulation see William R. Childs, *Trucking and the Public Interest: The Emergence of Federal Regulation, 1914–1940* (Knoxville: University of Tennessee Press, 1985).

[64] M. L. Ramsay, *Pyramids of Power: The Story of Roosevelt, Insull and the Utility Wars* (Indianapolis: The Bobbs–Merrill Company, 1937) and Stephen Raushenbush, *The Power Fight* (New York: New Republic, Inc., 1932). Also see McCraw, *TVA and the Power Fight, 1933–1939* for a study of public power during the presidency of Franklin Roosevelt. During presidential campaign speeches, he spoke in favor of developing four public projects to act as a "national yardstick" on power project operations. These were projects to be located on the St. Lawrence River, Muscle Shoals, Boulder Dam, and one on the Columbia River.

[65] D. B. Hardeman and Donald C. Bacon, *Rayburn: A Biography* (Austin: Texas Monthly Press, 1987), p. 172. See *Congressional Record*, January 4, 1935, pp. 116–18.

the government would impose regulations only where there was evidence of market failure. Judging from the history of Panhandle Eastern, there was ample justification for vigorous antitrust activity and for some controls on the financial practices of corporations. As it turned out, however, Congress would go substantially beyond those limits and create a new regulatory structure which would eventually burden the industry, consumers, and the national economy. In the 1930s, however, those problems seemed remote and the promise of public programs was compelling to a nation weathering the worst depression in the nation's history.

Congress enacted five major pieces of legislation that had a direct bearing on utilities. These included the Public Utility Holding Company Act of 1935 (PUHCA), the Federal Power Act of 1935 (FPA), the Natural Gas Act of 1938 (NGA), the Securities Act (1933), and the Securities Exchange Act (1934), which charged the Securities and Exchange Commission (SEC) with protecting those who bought and sold the securities of this industry and others.[66]

The Public Utility Holding Company Act, which occasioned "the greatest congressional battle in history," built on America's antitrust tradition.[67] Rayburn introduced a bill, House Resolution 5423, that included three titles: one dealing with holding companies; another with electric power; and a third with natural gas. Rayburn led the countercharge to the massive barrage of antiregulation propaganda which followed. The public utilities rained cloudbursts of telegrams, telephone calls, and threats of lawsuits on the legislators.[68] Rayburn

---

[66] Francis X. Welch, "Functions of the Federal Power Commission in Relation to the Securities and Exchange Commission," *George Washington Law Journal* vol. 46, no. 4 (1956), p. 81. For a discussion of the early history of the SEC see Joel Seligman, *The Transformation of Wall Street: A History of the Securities and Exchange Commission and Modern Corporate Finance* (Boston: Houghton Mifflin, 1982), pp. 1–100. For a general discussion of regulation in America see Thomas K. McCraw, "Regulation in America: A Review Article," *Business History Review*, 44, no. 2 (Summer 1975), pp. 159–83.

[67] D. B. Hardeman and Donald C. Bacon, *Rayburn: A Biography* (Austin: Texas Monthly Press, 1987), p. 167.

[68] For an interesting account of the struggle for the passage of the PUHCA see Hardeman and Bacon, *Rayburn: A Biography*, pp. 167–99.

was forced to compromise, and he shortly separated out the original version of the Natural Gas Act, Title III of House Resolution 5423. This version of the Act was not acceptable to a wide array of interested parties in part because it called for the FTC to regulate natural gas pipelines as common carriers. This issue embodied the "private vs. public" argument inherent in any regulatory effort: were pipelines private firms free to contract for gas purchases and sales or did pipelines exist to serve the public interest? It would, in fact, be three years before a successful compromise measure for natural gas could be guided through the legislature.[69]

Meanwhile, Rayburn and his colleagues concentrated on the Public Utility Holding Company Act, which mandated the abolition of all utility holding companies consisting of more than three tiers and restricted corporate organizations to a single, integrated utility system. The act gave the SEC discretion in dissolving the holding companies, but it specifically required utilities to separate their natural gas from their electric operations. The Act also required utility companies to register with the Securities and Exchange Commission and provide the agency with detailed financial information. Many of the large utilities ignored these rules until 1938, when the Supreme Court rejected a challenge by the Electric Bond & Share Company and upheld the legality of the PUHCA.[70]

---

[69] *Congressional Record*, House, 74th Cong., 1st sess., vol. 79, part 1, pp. 374–8 (January 11, 1935). During consideration of the Hepburn Act (1906), 34 U.S. Stat. 584, which imposed common carrier status on oil pipelines, the possibility of making natural gas pipelines common carriers was also considered. Congress decided not to impose common carrier status on gas pipelines because it was argued that development of the industry would then stop. See *Congressional Record*, May 4, 1906, pp. 6361 and 6371. Also see Bragdon, *The Federal Power Commission and the Regulation of Natural Gas*, pp. 29–30 and 59–63; and House Report, no. 764, 83rd Cong., 1st sess., July 10, 1953, "Why Natural Gas Pipelines are not Common Carriers." Significantly, the original version of the Natural Gas Act designated the Federal Trade Commission as the regulatory agency and imposed common carrier status on the natural gas pipeline industry. Rayburn himself, however, told his colleagues that natural gas pipelines were not by nature common carriers and that the structure of the industry precluded pipelines from being designated by the legislature as common carriers.

[70] Ellis W. Hawley, *The New Deal and the Problem of Monopoly*, p. 336.

The Natural Gas Act

During the three years following approval of the PUHCA, Congress debated several versions of the natural gas legislation.[71] As compromise legislation was developed, the Federal Power Commission (FPC) was designated as the new natural gas regulatory agency. Following the guidelines of state regulations and such federal agencies as the ICC, Congress decided to impose entry and price controls on the industry.[72] In the final form, the measure was "ambiguously worded, highly discretionary, and quite acceptable to both the regulated industry and the consuming public."[73] President Franklin D. Roosevelt signed the Natural Gas Act into law on June 21, 1938.

The FPC now had extensive power over interstate natural gas pipelines, as well as the interstate electric power business. The agency could approve and set "just and reasonable rates," extend the right to provide and abandon service, order improved facilities, and require extensive documentation of operations and finances. It could as well ascertain costs for rate making, suspend rates, and otherwise constrain and dictate the

[71] Congress originally created the FPC in 1920 (under the Water Power Act) to regulate the burgeoning water and hydroelectric power industries. In 1930, the FPC became an independent body with five full-time commissioners. Under the Federal Power Act of 1935, the FPC acquired the power to regulate the interstate transmission of electric power after President Roosevelt issued an executive order specifically addressing that unregulated industry. "Memorandum on Water Power Legislation," October 25, 1917. File Record Group-138, FPC, History of the Federal Power Act, 13–14, National Archives. Record Group hereafter referred to as RG. National Archives hereafter referred to as NARA. Also see John G. Clark, *Energy and the Federal Government: Fossil Fuel Policies, 1900–1946* (Urbana: University of Illinois Press, 1987), p. 280; and Sanders, *The Regulation of Natural Gas*, p. 66.

[72] For more information regarding the legislative history of the Natural Gas Act see Dozier A. DeVane, "Highlights of the Legislative History of the Federal Power Act of 1935 and the Natural Gas Act of 1938," *George Washington Law Journal* vol. 46, no. 4 (1956) and Bragdon, *The Federal Power Commission and the Regulation of Natural Gas*. Also see Kenneth Karl Marcus, *The National Government and the Natural Gas Industry*, pp. 110–34 and *Congressional Record*, vol. 79, part 2, pp. 1936, 1624.

[73] M. Elizabeth Sanders, *The Regulation of Natural Gas: Policy and Politics, 1938–1978* (Philadelphia: Temple University Press, 1981), p. 11.

operations of any interstate gas company.[74] In administering these powers, the FPC was required to conduct hearings on a case-by-case basis, using that evidence to decide to grant, or not to grant, a certificate of public convenience and necessity. A natural gas company had to have this certificate before it could engage in interstate commerce. With the certificate, a pipeline company held a federally sanctioned franchise over a particular service area. The federal government had in effect replaced a market managed by cartels with a political cartel managed by regulators.

Federal certification, it was hoped, would solve numerous problems which had plagued the industry. To obtain a certificate a pipeline had to have under contract an adequate supply of natural gas (usually for twenty years) appropriate to its market – earlier in the century some midwestern pipelines ran out of gas within only a few years. The line also had to have sound and proven financing and was allowed to charge only "just and reasonable" rates. In essence, the certificate became independent confirmation that the pipeline had excellent prospects, and indeed, ". . . certificates became a sine qua non for the sale of bonds financing new pipeline ventures."[75] In return for compliance with the certification procedures, the FPC allowed pipelines to set gas prices which reflected the line's costs of operation (including gas costs, and "rate base," or depreciated assets) plus an estimated rate of return of 5.7 to 6.5 percent. The FPC approved the customers and their respective allocations of gas.[76]

The immediate effect of the New Deal legislation on the structure of the utility industry seemed to be minimal. The Federal Power Commission urged the industry to cooperate because "immeasurable benefits from the enactment of the

[74] 52 U.S. 821 (1938).

[75] Sanders, *The Regulation of Natural Gas*, p. 50. Also see Richard W. Hooley, *Financing the Natural Gas Industry* (New York: Columbia University Press, 1961).

[76] The Natural Gas Act did not convey the right of eminent domain to natural gas pipelines. For the background of the 1947 eminent domain bill for natural gas pipelines, see Christopher James Castaneda, *Regulated Enterprise: Natural Gas Pipelines and Northeastern Markets, 1938–1954* (Columbus: Ohio State University Press, 1993), pp. 94–119.

Natural Gas Act will follow for the public and industry."[77] But at first it was not apparent what these benefits would be. Historian Richard Vietor has noted that "The Securities and Exchange Commission did not begin arranging divestitures of holding companies in any serious way until 1940. Company registrations, information gathering, and legal challenges accounted for much of the delay."[78] Ultimately, however, the SEC would investigate 929 gas and electric firms (1940–54) and would divest 158 of the former (valued at $874 million) and 259 of the latter (worth $9 billion).[79]

The industry would be decisively changed by these developments, but even before the divestments, the new public policy environment would provide William Maguire the opportunity he needed to recover control of the Panhandle Eastern pipeline. By breaking the hold of the cartel, New Deal regulation would in this instance actually encourage innovation. The cartel attempted to carefully control the introduction of natural gas into its established markets while maximizing its investment in manufactured gas. Under the new political economy, however, the large utilities would be on the defensive. Entrepreneurs like Maguire would be able to exploit that situation to their own advantage, mustering political power to best the economic power of their larger rivals. No longer would first movers have great advantages. No longer would corporate financial power alone be used to block entry and forestall change. In this new environment, the natural gas industry would flourish for several decades and Maguire would secure a major victory over his rivals.

[77] Clyde L. Seavey, "Federal Regulation of the Transportation and Sale of Natural Gas in Interstate Commerce," *American Gas Association Proceedings* (New York: AGA, 1938), pp. 126–9. On November 3, 1939, the FPC set forth a new standardized system of accounts to be effective on January 1, 1940, for all interstate natural gas companies. These new rules required companies to determine their system's original cost for rate filings. See Federal Power Commission, *Annual Report, 1939*, p. 4.

[78] Vietor, *Contrived Competition*, p. 100.

[79] Ibid.

4

The triumph of William Maguire

The new regulatory environment propelled by the depression-era political economy was a disaster for many businessmen, but it turned out to be very much to the liking of William Maguire. He would turn the antitrust and regulatory impulses of the New Deal to his advantage. He would in fact show himself to be a master competitor in this new arena. Competition did not end when political controls were imposed. Indeed, regulated competition forced businessmen to learn new rules. As would be apparent, William Maguire represented the new breed of entrepreneur in the utility business because he proved adept at comprehending the new political economy. Unlike Insull, Gossler, and Doherty, he had not led a large public utility firm during the era of unregulated interstate markets. He had not had the opportunity to do what the market allowed those men to do: develop new technologies in electric power and gas production, transmission, and distribution and deliver energy to the consumers *they chose* to serve. Having established powerful positions, those utility company captains used their power to keep out would-be entrepreneurs like both Maguire and Parish. In that sense, the free market had not been free to all – it had just been free of federal political interventions.

As of the mid-1930s, however, the market was no longer devoid of federal power. Parish had avoided federal prosecution for mail fraud charges, but he, like Maguire, was no longer a

Figure 4.1 William G. Maguire

member of the Panhandle board. Maguire, unlike Parish retained a strong position in the utility industry, and he soon
capitalized on a political environment hostile to giant enterprise. His first real break came as a result of the U.S. Department of Justice's interest in the Columbia Gas–Panhandle
Eastern dispute. On March 6, 1935, the U.S. Attorney General's office filed an antitrust suit against Columbia Gas in the
U.S. District Court in Wilmington, Delaware (Table 4.1). Attorney General Homer S. Cummings told reporters that the suit
was filed as part of an independent investigation and was not
directly related to the FTC investigation of the utility industry.[1] The suit alleged that Columbia Gas, Columbia Oil, and
certain officers and directors of the two companies had inter-
fered with Panhandle's business. The Department of Justice
also charged Columbia Gas and J. H. Hillman, Jr., with conspiring to restrain Mo–Kan from conducting business in the states
lying between Kansas and the Michigan–Ohio area. Soon after
Cummings filed the government's suit, Mo–Kan's receivers
filed a similar action against Columbia. William Maguire now
mounted a vigorous plan, to wrest control of Panhandle Eastern from Columbia Gas.[2]

Initially, it appeared that Maguire would once again strike
out. The Justice Department decided early in 1936 to accept a
consent decree by Columbia Gas and several other defendants
who would acknowledge their violation of the Sherman Antitrust Act. The consent decree stipulated that Columbia could
own Panhandle Eastern stock but could not control its operations. It also provided for the dissolution of the voting trust and
for the recapitalization of Panhandle Eastern. Columbia Gas
would still be able to determine when and how Panhandle expanded service, but now Columbia had to be careful that it did
not violate the new decree.[3]

---

[1] "Cummings on Columbia Gas Suit," *Wall Street Journal*, March 8, 1935.
[2] *U.S.A. v. Columbia Gas and Electric Corporation, Columbia Oil and Gasoline
Corporation, George H. Howard, Philip G. Gossler, Charles H. Munroe, George
W. Crawford, Thomas B. Gregory, Edward Reynolds, Jr., Burt R. Bay, and
John H. Hillman, Jr.* Motion for summary judgment and affidavit of Arthur
G. Logan in support thereof, with exhibits. Served February 16, 1942. No.
1099 In Equity, PECA.
[3] See "Decree Blocks Control over Pipeline Firm," *Denver Post*, January 31,
1936.

Table 4.1. *Columbia System: Principal active companies, December 31, 1937*

Amere Gas Utilities Company
Atlantic Seaboard Corporation
Binghamton Gas Works
Bracken County Gas Company
Central Kentucky Natural Gas Company
The Cincinnati Gas & Electric Company
Cincinnati Gas Transportation Company[a]
The Cincinnati, Newport, and Covington Railway Company
Columbia Corporation[a]
Cumberland and Allegheny Gas Company
The Dayton Power and Light Company
Fayette County Gas Company
Gettysburg Gas Corporation
Greensboro Gas Company
The Hamilton Service Company
The Harrison Electric and Water Company
Home Gas Company
Huntington Development and Gas Company
Huntington Gas Company
Indiana Gas Distribution Company
The Keystone Gas Company, Inc.
The Loveland Light and Water Company
Manufacturers Gas Company
The Manufacturers Light and Heat Company
Miami Power Corporation[a]
Michigan Gas Transmission Corporation
Natural Gas Company of West Virginia
The Northwestern Ohio Natural Gas Company
The Ohio Fuel Gas Company
Panhandle Eastern Pipe Line Company
Pennsylvania Fuel Supply Company
Point Pleasant Natural Gas Company
The Union Light, Heat and Power Company
United Fuel Gas Company
Virginia Gas Distribution Corporation
Virginia Gas Transmission Corporation
Warfield Natural Gas Company

[a] Nonoperating companies
Source: Albert F. Dawson, *Columbia System: A History* (New York: J. J. Little and Ives Company, 1938), p. 154.

Meanwhile, Parish and Maguire had formed separate Mo–Kan stockholders' committees in an effort to regain control of Panhandle. The pipeline was the crucial element in this struggle and Maguire was determined to acquire it. Parish's "Stockholder Committee of Chicago" was no match for Maguire's "Protective Committee for Stockholders," chaired by Robert W. Woolley of Washington, D.C. Woolley was a prominent member of Washington political circles. He had served as the chief investigator for the Stanley Commission's investigation of U.S. Steel during 1911 and 1912, and he had been a member of the Interstate Commerce Commission (ICC) from 1917 to 1921. Equally important to Maguire, Woolley was a close personal friend of Attorney General Cummings.[4] Cummings was important because Maguire's strategy now focused on Columbia's weak link, the consent decree. His "Protective Committee for Stockholders" accused Columbia of violating the decree by conspiring to gain control of Panhandle and to destroy Mo–Kan by refusing to contract for gas sales on its behalf.[5]

Under the decree, the federal government had direct oversight of the corporate organization and policies of Panhandle Eastern. It forced the firm to reorganize its board. The firm's president, Burt Bay, resigned, as did board members Charles Munroe and J. H. Hillman, Jr.[6] Columbia's 50-percent share of Panhandle stock was put in the name of Gano Dunn, an accomplished engineer and president of J. G. White Engineering Company of New York (Chart 4.1). Mo–Kan's receivers continued to control that organization's Panhandle stock. From the start, Dunn's involvement attracted Maguire's ire. He charged that Dunn was a mere agent of Columbia Gas, and some of the engineer's actions seemed to confirm this accusation.[7]

One of Dunn's first acts was to recapitalize Panhandle Eastern. Both Mo–Kan and Columbia Gas had owned 1,149 shares each of Panhandle Eastern stock. Through a stock dividend

---

[4] Eckles, C. III, p. 87, IHS.

[5] Panhandle Eastern Pipe Line, *Minutes*, February 4, 1936. Also see Sanders, *The Regulation of Natural Gas*, p. 33.

[6] Panhandle Eastern Pipe Line, *Minutes*, January 31, 1936. Burt Bay thereafter became president of Northern Natural.

[7] Gossler had once served as vice-president and director of the same J. G. White Engineering firm.

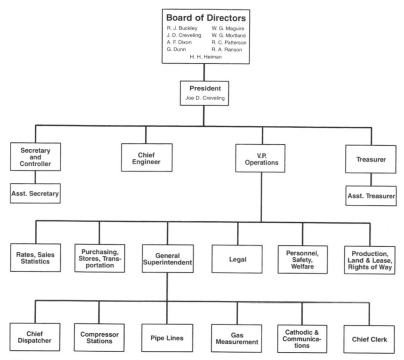

Chart 4.1 Panhandle Eastern organization chart (1937) prior to change of control

issue, these amounts were increased to 324,326 shares each. Dunn then issued an additional 160,000 shares of common, 80,000 of which were sold to Columbia. The remaining 80,000 were offered to Mo–Kan stockholders. Mo–Kan purchased 15,149 of the shares, and its stockholders bought the remainder. The result was an imbalance between Columbia Gas and Mo–Kan. Despite the consent decree, Columbia had increased its control over Panhandle Eastern.[8]

[8] Panhandle Eastern Pipe Line, *Minutes*, March 24, 1936 and May 25, 1936. Dunn also recapitalized a defaulted $11 million note by issuing to Columbia Gas $10 million in Class A preferred stock and $1 million in Class B preferred. The Class B stock carried with it the right to elect two additional board members, giving Columbia a clear majority of directors. Maguire vigorously but unsuccessfully opposed Columbia Gas's purchase of the Class B stock.

Expansion to Detroit

This development notwithstanding, Columbia felt compelled by the decree to expand the Panhandle Eastern system in order to demonstrate that it was not attempting to squelch the pipeline's business. The same day that Federal Judge John P. Nields issued the decree, Gossler announced Columbia's plans to build an extension connecting the Panhandle system to Detroit.[9] Gas sales to Citizens Energy of Indianapolis had not materialized, so the Detroit metropolitan market was important to the pipeline.

Detroit was relying almost exclusively on locally produced manufactured coal gas distributed through the Detroit City Gas Company; this firm was a subsidiary of United Light & Power Company, a large holding company controlled by financier Harrison Williams.[10] Natural gas had not penetrated this market in part because of an ongoing local debate regarding the city's intention to require the gas company to become a publicly owned municipal franchise, but the primary reason was the cost of the new energy source. For many years natural gas appeared to be more expensive than the city's existing supply of manufactured gas.[11]

By the mid-1930s, however, the City of Detroit decided that this was no longer the case and that the Detroit City Gas Company was misrepresenting its situation. New studies indicated that regional natural gas could be used in Detroit, contrary to the utility's contentions.[12] And in 1935, a Detroit City Council

[9] "Pipeline to be Laid by Columbia Gas," *New York Times*, January 31, 1936. Detroit did have access to natural gas in the late nineteenth century, but poor drilling methods and inefficient distribution led to a very quick dissipation of local reserves.

[10] See 39 FTC 47, 52, 65–6, 82–4. In 1908, the American Light & Traction Company of Chicago acquired three separate Detroit gas utilities and merged them into the Detroit City Gas Company. American Light & Traction was subsequently acquired by United Light and Power controlled by Cyrus Eaton, who later lost control in the 1920s to Harrison Williams.

[11] 83 FTC Report 132–3.

[12] In early October 1930, for example, the Seagraves Company, through its Northern Industrial Gas Company subsidiary, petitioned the Detroit City Council to sell up to 40 MMcf/d to Detroit at $.40 per Mcf. The following day, the Columbia Gas board agreed to build a line from Toledo to Detroit. Subsequently, Columbia decided to purchase the Seagraves interests for $10 million and never built either line to Detroit.

investigation revealed that Panhandle Eastern had sought a gas sales contract with the Detroit Company years earlier. This episode unfolded after Senator Gerald Nye denounced the utility industry and called for an investigation of the entire natural gas industry. As Nye proclaimed (December 4, 1934): "Whole populations of cities and states have been 'sold down the river' in the secret bargaining among the natural gas monopolists." Nye's charges led to a Senate Judiciary Committee investigation of the natural gas industry.[13]

Nye picked out Columbia Gas's manipulation of Panhandle Eastern for vituperation:

> It is a tale of fraud, of contract violation, commercial bribery, blackmail, stock market raiding, of bold division of territories among the barons, and of imprudent flouting of the antitrust laws. . . . This is as rich and colorful a yarn of modern commercial piracy as has come to my knowledge since Judge Landis broke up the old Standard Oil Company. The broad aspects of its revelations show that without Federal regulation, strict and sharp, the utilities can and to this day do dispose of competition in as ruthless a fashion as old John D. Rockefeller ever dreamed.[14]

In response to Senator Nye's statement, a newspaper reporter suggested that Detroit Mayor John W. Smith contact Frank Parish, who had advertised locally his interest in bringing natural gas to the city. The mayor did so, and Parish agreed to speak to the Detroit City Council about the fuel. William G. Woolfolk, President of the Detroit City Gas Company, appeared at this same meeting in February of 1935. This fifty-eight-year-old engineer had earlier attempted to purchase Panhandle Eastern for his utility, but now he was an opponent of natural gas.[15]

---

[13] The Cities Alliance, "Detroit Gets Natural Gas!," *The Natural Gas Monopoly* (Detroit: The Cities Alliance, n.d.), p. 5. Also see "Natural Gas," *Fortune* (August, 1940), p. 100.

[14] "Natural Gas," *Fortune* (August 1940), p. 100; and The Cities Alliance, *The Natural Gas Monopoly*, p. 13. Also see Eckles, C. VI, p. 4, IHS; and *Congressional Record* vol. 79, 3785 (March 16, 1935); p. 6510 (April 29, 1935).

[15] In 1932, Woolfolk, then a consulting engineer for the Detroit City Gas Company, sought to purchase for that utility a one-half interest in Panhandle Eastern; previously, Woolfolk had worked as a rate "trouble shooter" for Insull. Woolfolk was unsuccessful in acquiring an interest in Panhandle Eastern, but he soon thereafter became president of Detroit City Gas. See SEC Dockets 54-25; 59-11; 59-17; and Michigan Public Service Commission, Case D-3109, p. 334. In 1939, Woolfolk became chairman of the board of

Parish assured the council that natural gas could be transported economically to Detroit. He shocked the council members
by presenting a signed contract between Panhandle Eastern
and the Detroit City Gas Company. As the contract indicated
negotiations to bring natural gas to Detroit had begun at least
three years earlier. Woolfolk now admitted that, "Well, there
had been some negotiations" which had never been divulged to
the city government.[16] What he did not explain was that his firm
was trying to prevent natural gas from disrupting its existing
business or its business connections with other firms involved in
the production of coal and manufactured coal gas. As Woolfolk
knew, once introduced into a city, natural gas quickly overtook
and then supplanted manufactured gas. This had occurred recently in Chicago after the construction of NGPL, the "Chicago
Line."[17] Hoping to forestall a similar outcome, Woolfolk offered
Parish – no longer in any position of authority at Panhandle –
employment as a consultant, and Parish quickly accepted.[18]

The city council, however, was harder to co-opt. The council
put up for vote a proposal that the city transform Detroit City
Gas into a municipally owned franchise and allow the council
to control its rates. In addition, the city proposed to construct
its own "belt line" natural gas pipe system, which would circle
Detroit and provide natural gas to metropolitan industries.

This threat persuaded Columbia and Panhandle to sell natural gas to Detroit. But there was a catch. They would only do so
if they could sell gas not to the city but to Detroit City Gas. In
the meantime, the city attorney had informed the mayor's office that the proposal to municipalize Detroit City Gas required
the approval of three-fifths of the city's taxpayers (not just a
simple majority of registered voters). Before this issue could be
resolved, Detroit City Gas announced that it had signed a con-

United Light & Power Company and the American Light & Traction Company, two holding companies in the United Railways Company system that
controlled Detroit City Gas among many other firms. Also see *Congressional
Record*, vol. 79, p. 3785 (March 16, 1935) and vol. 79, p. 6510 (April 29,
1935).

[16] 83 FTC Report 131.

[17] For an account of the process through which Peoples Gas Light & Coke
Company began converting its service area from manufactured to natural
gas, see Stotz and Jamison, *History of the Gas Industry*, pp. 214–18.

[18] 83 FTC Report 130–1.

tract with Panhandle Eastern (dated August 31, 1935) for 90 MMcf/d of natural gas.[19]

Even this preemptive strike could not assuage Mayor Smith, who introduced a resolution in city council condemning the utility firms which had prevented Detroit from receiving natural gas in the past. Smith noted in the resolution that Columbia officials had already admitted having paid $10 million to purchase the Northern Industrial Gas Company to prevent it from transporting natural gas from West Virginia to Detroit. The mayor also alluded to a sworn statement confirming that Columbia Gas had used similar tactics when investing in Panhandle Eastern. Smith charged that the holding companies of Columbia Gas, Cities Service, and Standard Oil (N.J.) had conspired to control all the natural gas markets in the Midwest. This controversy prompted the U.S. Conference of Mayors to call for a publication to be distributed through the Cities Alliance detailing Detroit's experience with the natural gas industry. The Cities Alliance also called for regulation of the entire natural gas industry.[20]

Columbia Gas had reason to be concerned, but, conveniently, the new contract between Panhandle Eastern and Detroit City Gas stipulated that deliveries start by July 1, 1936. This date would make it difficult for the city to do much before the gas began to flow. Nevertheless, much had to be done in a short time to make that deadline. The cost of extending Panhandle Eastern's line to Detroit was estimated to be approximately $8 million, and Panhandle had no cash or financial resources. To move the venture along, Columbia Gas formed a new wholly owned subsidiary – the Michigan Gas Transmission Corporation – to construct and operate the line to Detroit. In the meantime, Columbia Oil acquired portions of Panhandle Eastern's system in Indiana, the Indiana Gas Distribution Corporation, and connected it with Michigan Gas. Columbia then owned all of Panhandle's system east of the Illinois–Indiana border including the only natural gas pipeline connecting with Detroit.

On March 17, 1936, Panhandle Eastern signed an agree-

19 83 FTC Report 134–5.
20 Ibid., p. 137. Also see 47 FTC Report on Columbia Gas & Electric Corporation (November 13, 1932); and 66 FTC Report on the Cities Service Company (June 4, 1934). Also see The Cities Alliance, *The Natural Gas Monopoly*, PECA.

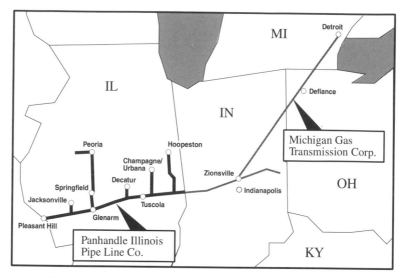

Map 4.1 Michigan Gas Transmission Corporation, 1936

ment to sell gas to Michigan Gas Transmission for resale to the Detroit City Gas Company. Columbia Gas at once launched plans to build the new line, originating at Panhandle Eastern's Zionsville, Indiana, compressor station. The twenty-two-inch line (Map 4.1) would extend 230 miles to a measuring station at the River Rouge plant of the Detroit City Gas Company.[21] Michigan Gas was completed quickly, and on July 9, 1936, it began delivering natural gas to Detroit. Table 4.2 shows that gas sales there gave Panhandle its first year of profits: approximately $1.7 million. In the previous year, the firm had lost more than $300,000.

Detroit City Gas now began a large-scale conversion of its distribution system from manufactured to natural gas. This involved adjusting the appliances of its gas customers to accept the higher BTU natural gas. Frank Parish, through his newly formed Universal Engineering Company, received a $150,000 contract, plus $3,000 per month, to begin converting Detroit's residential burners. Parish subsequently used $75,000 of this

---

[21] See Thomas Weymouth to Gano S. Dunn, February 15, 1936, Eckles Material, no. 9, PECA.

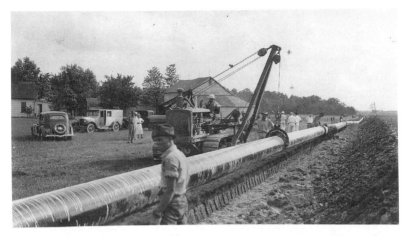

Figure 4.2 Lowering a section of the Michigan Gas Transmission line into the ditch, 1936

fee to buy natural gas reserves in southwest Kansas, some of which he later sold to Woolfolk. The conversion of Detroit City Gas's market area was followed by a similar conversion throughout the state of Michigan.[22]

These changes were all linked in one way or another to the decree, which precipitated the consolidation of the Panhandle system. In October 1936, Panhandle Eastern acquired all the assets of the Texas–Interstate Pipe Line Company. Later, during the summer of 1938, Panhandle Eastern also acquired all the mainline facilities of the Panhandle–Illinois Pipe Line Company. These actions placed all of the major transmission facili-

[22] Eckles, C. III, p. 82, IHS. For descriptions of manufactured to natural gas conversions in other cities see Nicholas B. Wainwright, *History of the Philadelphia Electric Company: 1881–1961* (Philadelphia: Philadelphia Electric Company, 1961), p. 319; David Grozier, "The Brooklyn Union Natural Gas Conversion: Biggest Changeover in the World," *Gas Age*, January 1, 1953, pp. 29–73; Joseph A. Pratt, *A Managerial History of Consolidated Edison, 1936–1981* (Consolidated Edison Company of New York, 1988), pp. 166–71; and (for Chicago) Stotz and Jamison, *History of the Gas Industry*, pp. 214–18. Parish sold a one-third interest in the Kansas properties to William Woolfolk for $50,000, and Woolfolk acquired ownership in forty-four Hugoton gas fields one half of which were owned by Panhandle Eastern.

Table 4.2. *Panhandle Eastern Pipe Line Company and subsidiary companies: Statement of Income 1931–1936*

	1931	1932	1933	1934	1935	1936
Gas Operating Revenue	$1,015,879	$1,901,451	$2,585,397	$3,048,518	$3,611,865	6,037,868
Operating expenses:						
Operation	442,142	910,304	888,273	920,668	978,335	1,380,665
Maintenance	32,556	155,071	110,832	88,901	86,074	122,872
Depreciation/Amortization	272,224	582,575	759,713	764,640	768,197	1,084,596
Taxes	31,969	219,460	262,375	270,204	284,598	590,294
Total expenses	778,891	1,867,410	2,021,193	2,044,413	2,117,204	3,178,427
Net operating revenues	236,988	34,041	564,204	1,004,105	1,494,661	2,859,441
Other income	2,497	28,238	8,235	4,055	3,012	5,671
Gross income	239,485	62,279	572,439	1,008,160	1,497,673	2,865,112
Interest & other deductions	552,127	1,865,250	1,874,555	1,880,167	1,823,727	1,153,927
Net Income or (loss)	⟨312,642⟩	⟨1,802,971⟩	⟨1,302,116⟩	⟨872,007⟩	⟨326,054⟩	1,711,185
Losses capitalized - Cr.	493,588	461,391	–	–	–	–
Balance to surplus	$ 180,846	$⟨1,341,580⟩	$⟨1,302,116⟩	$⟨872,007⟩	$⟨326,054⟩	$1,711,185

ties, from the producing fields in Texas to the Illinois–Indiana border, under the direct control of Panhandle. The Indiana facilities and Michigan Gas Transmission remained wholly owned subsidiaries of Columbia Gas, which clearly had no incentive to loosen its grip on Panhandle.[23]

The revival of Mo–Kan

By improving Panhandle Eastern's revenues, however, Columbia had actually bolstered Mo–Kan's balance sheet. As a result, the Delaware Chancellor's Court terminated most of the duties of Mo–Kan's receivers on October 15, 1937. Mo–Kan's reactivated board elected William Maguire as president, and the firm regained direct control over its Panhandle stock. Frank Parish was now persona non grata at the revitalized Mo–Kan, where he had no authority. He had already lost his influence at Panhandle, and this enabled Maguire to pursue his own personal strategies as he struggled to acquire complete control of Panhandle from Columbia Gas.[24]

As they revived Mo–Kan, William Maguire and Faison Dixon occupied the seats on Panhandle Eastern's board that had been held by Mo–Kan's lame duck representatives. Maguire immediately requested that Panhandle open all its records to Dixon's engineering firm, Brokaw, Dixon & McKee, so that the consulting firm could produce a full report on the pipeline. Maguire sought all of the detailed information he could get on Panhandle's operations.[25]

Maguire insisted that Columbia was still illegally dominating Panhandle's affairs. He cited its control of Panhandle's Indiana system, the "robbery" of the Michigan Gas Transmission line, and the unfair recapitalization plan. He was not crying in

---

[23] The distribution related facilities were merged into another newly created subsidiary, the Illinois Natural Gas Company. Panhandle Eastern, somewhat freed from Columbia's overt control, signed a fifteen-year gas sales contract with Kentucky Natural Gas, for 2.5 MMcf/d increasing to 10 MMcf/d by the sixth year of the contract. Deliveries were to begin on September 1, 1938.

[24] Frank Parish, interview by Robert Eckles, July 7, 1957, PECA. Parish later recalled that "my activities ceased at the time the Chancellor approved the settlement, and the stockholders elected new directors, and the receivership was lifted. Maguire won the proxy fight. I was out. . . ."

[25] Panhandle Eastern Pipe Line, *Minutes*, September 22, 1937.

the dark; indeed, these charges gave him the ammunition he needed to lobby the Department of Justice to reopen the consent decree. Maguire wanted complete control of Panhandle Eastern and to achieve that goal, he had to force Columbia Gas to divest the pipeline firm. He got the break he needed after Thurman Arnold, a foe of monopoly in all of its forms, became Assistant Attorney General in charge of Antitrust in March 1938. Arnold launched a new complaint (issued by Attorney General Murphy) charging Columbia Gas with violating the consent decree. Arnold discussed these matters with Gano Dunn during May of 1938.[26]

Although Arnold oversaw the complaint, he was not certain that Columbia had actually violated the decree. He had previously concluded that the consent decree was a practical solution to the problem and that perhaps it was enough. His lack of verve evoked a sharply worded response from the vitriolic Chicago *Tribune*:

> The [Assistant] Attorney General does not insist that the control of the Panhandle company by Columbia Gas is illegal and must be stopped. He says it is illegal only if people he doesn't like are in command. The possibilities of unjust discrimination and corruption in such a scheme are infinite. Here is another point at which the New Deal and the Nazi regime are fond of occupying identical positions.[27]

Maguire was less inclined than the *Tribune* to deal in hyperbole, but he now prepared to take Columbia to court. During the summer of 1938, Mo–Kan prepared a law suit charging Columbia with violating the consent decree. Mo–Kan wanted Columbia to be charged with contempt of court, to transfer Michigan Gas to Panhandle, to equalize the stock ownership, and then to divest itself of all its remaining Panhandle Eastern stock.[28]

The Justice Department asked Mo–Kan to wait before filing its petition so the government could:

[26] Mo–Kan, *Minutes*, July 14, 1938, PECA. Also see Panhandle Eastern, *Minutes*, May 31, 1938. Thurman Arnold served as Assistant Attorney General in charge of Antitrust from March 7, 1938 through March 16, 1943.

[27] *Chicago Tribune*, April 27, 1940. Also see "Natural Gas," *Fortune* (August 1940), p. 100.

[28] Mo–Kan, *Minutes*, July 14, 1938, PECA.

. . . use every effort to attain the ends sought by Mo–Kan's petition by negotiation, or if negotiation failed, by litigation. Such
ends included a modification of the decree to compel the sale of
Michigan Gas to Panhandle Eastern, the elimination of the additional 160,000 share issue of Panhandle common, and the divest-
ment of the remainder of the Columbia holdings.[29]

Now Maguire had a strong hand to play. In late 1938 he
received some additional support when a frustrated court indicated its continuing displeasure with Columbia's manipulation
of Panhandle and concluded that "the only effective way to
restore and maintain a position of free and independent action
for Panhandle Eastern is to require Columbia Gas to divest
itself of all stock of any class having existing or potential voting rights in Columbia Oil . . . [or] Panhandle Eastern."[30] Maguire now was very close to achieving his goal.

The following year the Justice Department filed a second
and then a third supplemental complaint against Columbia,
including a new charge that the firm had forced Panhandle to
allow it to build and independently operate Michigan Gas
Transmission company. It had done so, Justice said, even
though Panhandle Eastern had actually contracted for gas
sales to the Detroit City Gas Company. After reopening the
consent decree, the Justice Department also charged that Columbia prevented Mo–Kan from receiving natural gas from
Panhandle, that Columbia had never purchased natural gas
from the pipeline, and that Columbia "neglected opportunities
for Panhandle Eastern to sell natural gas to cities of Indianapolis, Ind.; Kansas City, Mo.; St. Louis, Mo.; Detroit, Michigan;
thereby forcing Mokan into receivership."[31] The government

---

[29] Ibid.

[30] Brief and argument of certain objectors to the proposed settlement in the
Court of Chancery of the State of Delaware. RE: Missouri–Kansas Pipe Line
Company, Frank P. Parish & Co., Kentucky Natural Gas Company, and
Indiana–Kentucky Natural Gas Corporation, PECA.

[31] See Mo–Kan, *Minutes*, February 10, 1940, PECA; and *U.S.A. v. Columbia
Gas and Electric Corporation, Columbia Oil and Gasoline Corporation,
George H. Howard, Philip G. Gossler, Charles H. Munroe, George W. Crawford, Thomas B. Gregory, Edward Reynolds, Jr., Burt R. Bay, and John H.
Hillman, Jr.* Motion for summary judgment and affidavit of Arthur G. Logan
in support thereof, with exhibits (February 6, 1939). Served February 16,
1942. No. 1099 In Equity, PECA.

also raided Gano Dunn's office, finding documents suggesting that Columbia officials had bribed a Justice Department attorney and had exercised undue influence on Dunn.[32]

Columbia Gas, with its back to the wall, began to negotiate with the Justice Department. During July 1939, the company submitted to the Delaware Court an "Exchange Plan" providing for the dissolution of Columbia Oil, the sale of Columbia Oil's $10 million of Class A preferred stock in Panhandle, and the sale to Panhandle of the Michigan Gas and Indiana Gas Distribution Company. Maguire was on the brink of victory, but then a Columbia Gas stockholder claimed the plan was unfair and delayed its implementation for two years. Meanwhile, the outbreak of war in Europe shifted attention in Washington from antitrust to efforts to maximize the nation's energy production. Maguire would have to wait.

Maguire scores – at last

Maguire stayed busy, looking for a firm to help him purchase the Panhandle Eastern stock owned by Columbia. Maguire found the business associate he needed in Kenneth S. Adams, chairman and president of the Bartlesville-based Phillips Petroleum Company. Adams knew that the days of Columbia's ownership of Panhandle were coming to an end; he saw an opportunity for his firm, which had become one of the largest natural gas producers in the country, to tap new markets. In early August 1939, Adams wrote to the Panhandle board inquiring about the possibility of purchasing Columbia's share of the stock and of selling Panhandle casinghead gas – gas produced in the process of producing oil – in the Texas Panhandle.[33] In New York, Maguire subsequently approached Hy Byrd, vice president and assistant secretary of Phillips, looking for support.[34]

He needed it by June 1941, when Maguire rode the Justice Department's support into a seven point agreement calling for Columbia to retire its holdings of Panhandle preferred, sell all

---

[32] *New York World Telegram*, March 14–15, 1939, PECA. Also see Eckles, Pt. I, C. VII, p. 17, IHS.

[33] Mo–Kan, *Minutes*, August 10, 1939; September 15, 1939; November 7, 1940; and January 7, 1941, PECA. Also see, Eckles, Pt. I, C. VII, p. 18, IHS.

[34] Phillips already had a business relationship with Panhandle Eastern which supplied natural gas to Phillips' facilities in Kansas and Missouri. Panhandle Eastern, *Minutes*, February 16, 1932 and November 6, 1935.

Table 4.3. *Natural gas sales by utilities in Michigan, 1933–1953 (millions of therms)*

Year	Natural gas	Manufactured gas
1933	5.7	148.0
1934	14.1	154.6
1935	28.4	161.3
1936	99.0	128.3
1937	239.0	44.7
1938	232.4	41.2
1939	269.3	43.2
1940	325.2	44.3
1941	356.8	49.4
1942	410.9	40.2
1943	516.9	19.8
1944	533.5	18.5
1945	549.8	16.9
1946	667.2	14.2
1947	799.5	13.6
1948	749.9	14.3
1949	797.3	12.9
1950	1,244.6	9.5
1951	1,474.8	3.0
1952	1,575.0	1.3
1953	1,674.6	1.0

Note: 1 therm is equivalent to 100,000 Btu.
Source: American Gas Association, *Historical Statistics of the Gas Industry, 1961* (New York: American Gas Association, 1961), pp. 254–5.

the stock of the Michigan Gas Transmission to Panhandle, allow Mo–Kan to create a Chairman of the Board position, and sell all of its Panhandle common stock. As Maguire informed his board triumphantly "it gave MoKan all of the objectives for which it has been contending since [I] was elected President [of MoKan]."[35] On February 6, 1942, Panhandle Eastern formally purchased from Columbia the profitable Michigan Gas Transmission Corporation (Table 4.3) and the Indiana Gas Distribution Corporation and began the process of integrating these once separate companies fully into its own system.[36]

Columbia moved more slowly on completing the rest of the

[35] Mo–Kan, *Minutes*, June 7, 1941, PECA.
[36] Panhandle Eastern, *Annual Report, 1941*, p. 5.

terms, giving Maguire additional time to acquire financial support.[37] He continued negotiations with Phillips Petroleum Company, which wanted an outlet for its gas but was leery of acquiring control of the pipeline and possibly coming under FPC regulation. Phillips preferred instead a joint purchase that would give Mo–Kan control, enabling it to expand into Ohio, Michigan, Indiana, and Missouri. Phillips considered dedicating as much as 175,000 acres of gas reserves in the Hugoton field to this joint project.[38] In March 1942, the SEC made that possible by officially ordering Columbia to divest its Panhandle Eastern holdings.[39]

By the summer of 1942, Maguire was on the edge of a complete victory. On July 6, 1942, Phillips signed a contract with Panhandle Eastern dedicating the Hugoton gas production to the pipeline; the contract set a price of $.04 per Mcf for five years with a $.005 per Mcf escalator every five years. The following day, Phillips and Panhandle Eastern signed an agreement providing for the purchase by Phillips of all of the Panhandle stock owned by Columbia. Phillips would buy 404,326 shares at a price of $26.12. Phillips agreed that it would sell half of these shares (202,163) to Mo–Kan.[40] Noting the plan would extricate "the companies and other interested persons involved from the problems which they face under the antitrust laws, the Holding Company Act, and a complex tangle of private litigation," the SEC approved the transaction on November 10, 1942.[41] As the *New York Times* observed:

> One of the bitterest battles in corporate history drew to a close
> yesterday when Columbia Oil & Gasoline Corporation, a subsid-

---

[37] Mo–Kan, *Minutes*, March 18, 1942, PECA. In the meantime, Mo–Kan settled its long-term antitrust lawsuits against Cities Service and Standard Oil. Although Mo–Kan at one time sought a multimillion dollar judgment against the two firms, it accepted an out-of-court settlement of $200,000.

[38] Mo–Kan, *Minutes*, May 23, 1942, PECA.

[39] The SEC ordered cancellation of the Class B preferred as of March 3, 1942 and the Class A preferred was retired on February 6, 1942.

[40] "Gas Contract Between Phillips Petroleum Company and Panhandle Eastern Pipe Line Company" in Mo–Kan, *Minutes*, July 6, 1942, PECA. Also see Securities and Exchange Commission (October 2, 1942), pp. 229. Eckles, Pt. I, C. VII, p. 30, IHS.

[41] SEC, "Columbia Gas & Electric Corporation, et al.," no. 1642, October 2, 1942, p. 226. Also see *New York Times*, November 10, 1942 and December 9, 1942.

iary of the Columbia Gas & Electric Corporation, announced that it would relinquish control over the rich Panhandle Eastern Pipe Line Company.[42]

The end game

Although Maguire had victory at hand, he still found it necessary to remain patient as he closed in on the control of Panhandle. The sale of Panhandle stock to Phillips took several months and required several extensions of the agreement. Frank Parish surfaced to proclaim his opposition and to take his objections to the Department of Justice. The Department ignored Parish, but the necessary financial and legal work caused further delays.[43] Maguire wanted the deal to go through before Panhandle's next annual meeting (March 31, 1943), and March 30 was set as the date for closing. Phillips borrowed $5,750,000 from the Insurance Company of North America to acquire its 202,163 shares.[44]

Columbia Gas, however, continued to drag its feet. As a Phillips attorney recalled, Columbia's representatives "did everything they could up to the very last moment to prevent the consummation of the deal."[45] About $9 million was required at the closing, and Columbia demanded that Phillips pay the full amount in cash. This was an unusual request on a deal of this magnitude, but Phillips arranged for an armored car to deliver the cash to Columbia's bank. Columbia management then decided that a check would be satisfactory after all.

Still, at the closing, the Columbia representatives were "as rude and unmannerly as they could possibly be. . . ." According to the Phillips attorney, "they refused to talk to [Maguire] directly and would only talk to him . . . and his representatives through the Phillips men."[46] Despite bad manners, the deal went forward.

---

[42] *New York Times*, July 11, 1942.

[43] See Mo–Kan, *Minutes*, October 27, 1943 and December 15, 1943, PECA. Mo–Kan had earlier contracted former Senator Robert J. Bulkley (D–OH), to assist the firm on legal and regulatory issues in regard to the Panhandle Eastern–Columbia Gas situation; Bulkley received at least $25,000 for these efforts.

[44] "Dissolution Plan Laid to Lawsuits," *New York Times*, November 5, 1943.

[45] Unnamed Phillips Petroleum Company representatives, interviewed by Mr. Eckles, March 7, 1957, PECA.

[46] Ibid.

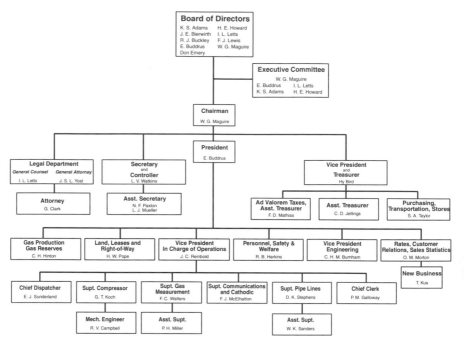

Chart 4.2 Panhandle Eastern organizational structure after regaining independence

Mo–Kan now owned 66 percent of Panhandle Eastern's common stock; Phillips owned 25 percent, and 9 percent was publicly held.[47] At the next Panhandle Eastern board meeting (April 5, 1943), Maguire replaced all but one of the Panhandle Eastern directors with his own supporters.[48] The newly elected board created a new position, Chairman of the Board, and Maguire, the victor, assumed that position of leadership (Chart 4.2).

Maguire quickly consolidated his position by establishing a new management team, and issuing a call for cooperation

[47] "Panhandle Eastern under New Control," *Panhandle Lines*, August 1943, p. 3. Also on the day of the closing, the controversial Class B preferred stock was retired; the Class A preferred had previously been retired.

[48] Panhandle Eastern, *Minutes*, April 5, 1943. Also see "Panhandle Eastern under New Control," *Panhandle Lines*, August 1943, p. 3; and Michael Wallis, *Oil Man: The Story of Frank Phillips and the Birth of Phillips Petroleum* (Doubleday, 1988); and Eckles, Pt. I, C. VII, p. 32, IHS.

through a new company publication, *Panhandle Lines*.[49] The first issue (August 1, 1943) proclaimed:

> The magazine is more than a medium for exchanging news and views of the departments. It is a call to teamwork. With the recent transfer of control, some changes in the executive positions were inevitable; otherwise, personnel security and advancement are, as they should be, entirely based on loyalty, integrity, and hard work.[50]

Meanwhile, he was trying to bring under control a dissident group of stockholders who had formed a committee called the Stockholders of Missouri–Kansas Pipe Line Company. The committee accused Maguire of attempting to create a "dynasty" and of enriching himself at the stockholders' expense.

Maguire could not ignore this challenge to his authority. The committee included A. Faison Dixon and Arthur Logan, both of whom had recently supported Maguire. Dixon, then a consulting engineer for the firm, was a vice-president and director of Mo–Kan, while Logan had been counsel to the original Mo–Kan receivers and then for Mo–Kan from 1940 through September 1943. According to the committee, Maguire had received $198,140 in extra compensation during 1943, when the 16,000 Mo–Kan stockholders were receiving only $163,745 in dividends. Now, the committee charged, Maguire was planning to dissolve Mo–Kan, possibly leaving the stockholders with nothing. At the time, Mo–Kan's assets consisted almost en-

[49] Panhandle Eastern, *Minutes*, April 5, 1943. Panhandle Eastern's affiliation with Phillips brought a new group of executives to the pipeline firm. Edward Buddrus, manager of Phillips' natural gas and natural gasoline department became Panhandle's president. Soon after joining Panhandle, Buddrus became the first president of the newly organized Independent Natural Gas Association of America. Buddrus had worked for Phillips since the mid-1920s and supervised its purchase and sales of natural gas, carbon black, and gasoline plant operations. Hy Byrd, a vice-president and assistant secretary of Phillips became Panhandle's vice-president and treasurer. In addition, two other Phillips executives including Kenneth S. Adams, its chairman and president, and Don Emery, vice-president and general councel, became Panhandle directors. Maguire had been impressed with the character of John E. Bierwirth, who served on the Columbia Gas board and chose him as the only director appointed by Gano Dunn to be retained on the new Panhandle board.

[50] "The Forward View," *Panhandle Lines*, August 1, 1943.

tirely of Panhandle Eastern and Kentucky Natural Gas Company shares.[51]

Maguire responded to these charges with a powerful message to the stockholders, and his own two-part plan to reorganize the relationship between the two firms. First, he would eliminate Mo–Kan's debt by selling up to 163,710 shares of Panhandle Eastern stock. Even at $30 per share, $10 less than the current market price, that sale would raise almost enough to pay off the $5 million debt from the original purchase of the stock from Columbia Gas. Maguire implemented this plan and by the spring of 1944, the stock sale had netted $4.6 million.[52]

The second phase provided stockholders with the opportunity to exchange their Mo–Kan stock for Panhandle Eastern shares at a ratio of 9 (Mo–Kan) for 2 (Panhandle). Maguire did not personally participate in this "Mo–Kan Plan." By remaining one of the few substantial holders of Mo–Kan stock, he could personally control that firm's sizeable holdings of Panhandle shares.[53] Parish, who realized that Maguire was on the verge of achieving control of both companies, mounted one last-ditch campaign. Working with a Columbia Gas attorney, he tried to organize a new stockholders' group. He mailed $.01 post cards to Mo–Kan stockholders warning them of Maguire's bad management, and proposing a different plan.[54] A more formidable threat came from the European utility firm, Sofina, whose representatives began purchasing large quantities of Panhandle Eastern stock.

Maguire was able to survive all of these latter day challenges. He was helped in the case of Sofina by the U.S. Treasury Department. After Treasury moved to freeze Farben In-

---

[51] See Howard Butcher III, A. Faison Dixon, and J. Walter Taylor to the Stockholders of Missouri–Kansas Pipe Line Company, December 27, 1943; "MoKan's President Assailed by Group," *New York Times* (no date), MoKan file, PECA. See *New York Herald Tribune*, December 30, 1943. Under the heading "Mo–Kan cannot afford Maguire," the committee sent several notices to Mo–Kan stockholders requesting proxies for the upcoming Panhandle Eastern annual meeting.

[52] Board of Directors of Missouri–Kansas Pipe Line Company to Stockholders of Missouri–Kansas Pipeline Company, February 15, 1944, PECA.

[53] Ray Shibley, interview by Christopher J. Castaneda and Clarance M. Smith, January 13, 1994, PECA.

[54] Frank P. Parish to Stockholders of Missouri–Kansas Pipe Line Company, March 2, 1944, PECA.

dustrie funds in the U.S., the foreign purchases of Panhandle stock suddenly stopped.[55] He ignored Parish and countered the stockholders' committee by sending all of the stockholders selected statements Dixon and Logan previously had made to the board in support of Maguire. As the board minutes revealed, they had credited Maguire with the major role in Mo–Kan's success. Maguire also tried to justify the monies he had received from Mo–Kan and Panhandle Eastern during the extended battle against Columbia Gas.

In the end, Maguire won.[56] By late 1944, he had fended off these attacks and acquired complete control of Panhandle Eastern. When Maguire learned that Phillips Petroleum wanted to sell most of its 202,163 shares of Panhandle, he and Mo–Kan purchased 150,000 of these shares (February 9, 1945). Mo–Kan subsequently sold more than 200,000 Panhandle shares directly to its own stockholders. Maguire's control was at last complete.[57]

[55] See *Congressional Record*, June 19, 1940, pp. 12989–91; June 20, 1940, p. 13121; vol. 87, no. 206, November 18, 1946, p. A5517 and vol. 87, no. 210, November 25, 1941, p. 5624.

[56] William G. Maguire to the Stockholders of Missouri–Kansas Pipe Line Company, March 1, 1944, PECA.

[57] See, Michael Wallis, *Oil Man: The Story of Frank Phillips and the Birth of Phillips Petroleum* (New York: Doubleday, 1988); and Charles Hinton, interview by R. B. Eckles, February 13, 1956, PECA. Phillips's decision to sell its Panhandle Eastern stock was based at least in part on its decision to participate in the competitive proposed Michigan–Wisconsin Pipe Line. That line would be owned by MichCon and sell gas to it and MichCon's other gas distributor in Wisconsin. A Panhandle Eastern employee recalled that Phillips "disposed of the Panhandle stock and got off the Panhandle Board because negotiations had then reached the point on what is now known as the Michigan–Wisconsin Pipeline where it was no longer quite the thing to do to have two members on Panhandle's board when a competitive line was being constructed." Initially, Phillips suggested that the sale be made directly to Maguire and Byrd with each share priced at $50. Maguire also announced that Mo–Kan had sold the common and preferred stock of the Kentucky Natural Gas Corporation for $264,500 to Mr. J. H. Hillman, Jr., who then acquired the Memphis Natural Gas Company. Hillman then organized the Texas Gas Transmission Corporation on December 7, 1945, and merged his two gas firms into the newly created one. Texas Gas served customers in Indiana, Illinois, and Kentucky with gas produced in Louisiana and Texas. L. L. Waters, *Energy to Move . . . Texas Gas Transmission Corporation* (Texas Gas Transmission Corporation, 1985), pp. 30–5. William G. Maguire to

Political entrepreneurship

Maguire's triumph was the result of his persistence and his skill at political entrepreneurship.[58] The political economy of the industry prior to the New Deal had benefited most those entrepreneurs who could develop, in the words of business historian Alfred Chandler, "first-mover" advantages.[59] Chandler's concept of these advantages stressed only economies of scale and scope, but clearly the first generation utility moguls had acquired political and economic power that they used to counter aspiring entrepreneurs like Maguire. Insull, Doherty, Gossler, and Payne had grown up in the burgeoning utility business. They had been able to take advantage of entrepreneurial opportunities in a wide open developing industry which produced energy for an anxious public. Maguire entered that industry too late to be a first-mover, and he could not overcome the financial and political advantages of the utility captains. This was especially the case when the leaders acted in collusion to thwart entry. The cartel was powerful, and it used its power to protect its markets, even those being served by an outdated product like manufactured gas.

The New Deal strengthened the antitrust tradition in the American economy while imposing additional regulations to manage the interstate operations of this and other industries. It was not at all clear that additional rate-of-return regulations were needed to solve the problems of the natural gas industry; indeed, the market-failure test would suggest it was not. But antitrust worked in this case. In a political system prepared to break up large utility combines, Maguire identified an opportunity to encourage the Justice Department to force the separation of Panhandle Eastern from Columbia Gas. Patiently pressing

Stockholders of Missouri–Kansas Pipe Line Company, November 24, 1944, PECA.

[58] Jameson W. Doig and Erwin C. Hargrove, ed., *Leadership and Innovation: Entrepreneurs in Government* (Baltimore: The Johns Hopkins University Press, 1990). Doig and Hargrove describe political entrepreneurs as public servants who behave in the public interest as entrepreneurs. In this manuscript, political entrepreneur refers to an entrepreneur who is adept at manipulating the political and regulatory system for private gain.

[59] Chandler, *Scale and Scope*, p. 34.

his case and protecting his corporate flanks, Maguire finally broke the hold of the cartel.

During this epoch in Panhandle Eastern's history, Maguire succeeded while both Columbia and Frank Parish failed. Columbia Gas lost because its advantage of scale became a disadvantage under a reinvigorated antitrust policy. By responding more quickly to the new political environment, Columbia might have been able to preserve more of its position. But like many large companies accustomed to great power and profit, it fought a holding action instead. Parish lost because he had been unable to sustain a long-term, concerted effort to protect his financial interest in the pipeline firm. He had throughout been more concerned with short-term profits than long-term development; he later abandoned the gas industry to operate a dairy farm in Maryland.[60] Maguire's success was based on a long-term strategy which postulated that economic success now depended on success in adapting to the new role the government was playing in this and other industries. The reward for his correct perception and his ability to implement his strategy was an opportunity to show that he could administer the pipeline operations he had acquired as well as he had performed as a political entrepreneur.

[60] "Boy Wizard of Finance Recalls Wealthy Years," *The Kansas City Times,* December 14, 1963.

PART II

Managing regulation: The challenge

5

Competing in new forums

As Maguire and Panhandle Eastern entered the new era of independent operations, they were confronted by a number of complex competitive and regulatory situations, some of which had deep historical roots. One element of Panhandle's competitive strategy involved the Detroit market, where the firm sold its gas to the Detroit City Gas Company (later renamed Michigan Consolidated Gas Company, or MichCon). The SEC was in the process of breaking this corporation, among many others, out of its holding company structure (Chart 5.1) as part of Congress's mandate to simplify the nation's utility industry.[1] William Woolfolk, who headed MichCon's top holding company, could not prevent the breakup of his combine, but he believed that he could save the profitable gas distribution properties in Michigan and Wisconsin by linking them with an affiliated pipeline (Map 5.1). Despite official SEC disapproval, he set out to build a pipeline from the Hugoton field to serve the Wiscon-

[1] SEC ordered the dissolution of the United Light and Power Company under Section 11 (b) 2 of the PUHCA, 15 U.S.C. §79K(b)2. On March 20, 1941, the SEC issued the actual dissolution order; United Light and Power ceased to exist after July 1, 1945. The SEC ordered the dissolution of the second holding company, the United Light and Railway Company on August 5, 1941. This order also required the American Light & Traction Company to divest its nonintegrated holdings. See 9 SEC 833.

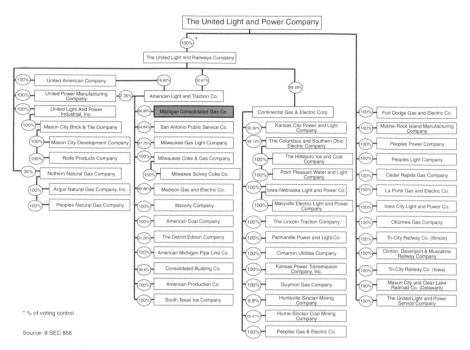

Chart 5.1 United Light and Power System, December 31, 1939

The United Light and Power Company

— (100%) * **The United Light and Railways Company**

- (100%) United American Company
- (100%) United Power Manufacturing Company
- (100%) United Light And Power Industrial, Inc.
- (100%) Mason City Brick & Tile Company
- (100%) Mason City Development Company
- (100%) Rolfe Products Company
- (35%) Nothern Natural Gas Company
- (100%) Argus Natural Gas Company, Inc.
- (100%) Peoples Natural Gas Company

— (16.92%) (32.67%) (2.35%) **American Light and Traction Co.**

- (99.99%) Michigan Consolidated Gas Co.
- (64.84%) San Antonio Public Service Co.
- (97.75%) Milwaukee Gas Light Company
- (100%) Milwaukee Coke & Gas Company
- (100%) Milwakee Solvay Coke Co.
- (99.98%) Madison Gas and Electric Co.
- (100%) Waverly Company
- (100%) American Coal Company
- (20.26%) The Detroit Edison Company
- (100%) American Michigan Pipe Line Co.
- (99.8%) Consolidated Building Co.
- (100%) American Production Co.
- (100%) South Texas Ice Company

— (99.59%) **Continental Gas & Electric Corp.**

- (92.92%) Kansas City Power and Light Company
- (99.12%) The Columbus and Southern Ohio Electric Company
- (100%) The Hillsboro Ice and Coal Company
- (100%) Point Pleasant Water and Light Company
- (100%) Iowa-Nebraska Light and Power Co.
- (100%) Maryville Electric Light and Power Company
- (100%) The Lincoln Traction Company
- (100%) Panhandle Power and Light Co.
- (100%) Cimarron Utilities Company
- (100%) Kansas Power Transmission Company, Inc.
- (100%) Guymon Gas Company
- (9.9%) Huntsville Sinclair Mining Company
- (25.47%) Hume-Sinclair Coal Mining Company
- (100%) Peoples Gas & Electric Co.

- (100%) Fort Dodge Gas and Electric Co.
- (100%) Moline-Rock Island Manufacturing Company
- (100%) Peoples Power Company
- (100%) Peoples Light Company
- (100%) Cedar Rapids Gas Company
- (100%) La Porte Gas and Electric Co.
- (100%) Iowa City Light and Power Co.
- (100%) Ottumwa Gas Company
- (100%) Tri-City Railway Co. (Illinois)
- (100%) Clinton, Davenport & Muscatine Railway Company
- (100%) Tri-City Railway Co. (Iowa)
- (100%) Mason City and Clear Lake Railroad Co. (Delaware)
- (100%) The United Light and Power Service Company

* % of voting control

Source: 9 SEC 856

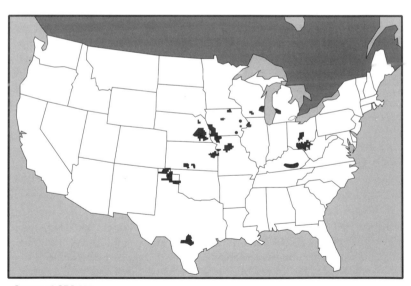

Source: 9 SEC 857

Map 5.1 United Light and Power Company: Areas served by utility subsidiaries

sin and Michigan gas distributors, including MichCon.[2] In the fall of 1944, Woolfolk announced the proposed line, the Michigan–Wisconsin Pipeline Company (Mich–Wisc). If the FPC certified the line and the SEC decided to leave this part of the holding company structure intact, Maguire would have a major competitive problem.[3]

Woolfolk not only wanted to create an integrated system, he wanted Mich–Wisc to replace Panhandle Eastern's gas deliveries to Detroit. The Mich–Wisc line would receive its entire gas supply from Phillips Petroleum Company, which dedicated 633,460 acres of reserves to Mich–Wisc.[4] This had prompted Phillips's to divest its one-fourth interest in Panhandle Eastern. Although Maguire then acquired complete control over Panhandle, he faced a significant competitor in the Mich–Wisc line, which was positioned to become the primary gas supplier to Detroit and MichCon.[5]

Maguire set out to turn back this challenge, exploiting the new political economy of natural gas as much as he had in acquiring control of Panhandle. This time, however, Maguire was the opponent of innovation, manipulating the regulatory system in an effort to prevent change much as Columbia Gas had done in the earlier struggles. Maguire had some important allies. Although MichCon may have expected the support of

---

[2] *Detroit News*, February 17, 1952. Also see, *Fuel Investigation*, Hearing before the Committee on Interstate and Foreign Commerce House of Representatives, 80th Cong., 2d sess., (Washington: GPO, 1948), pp. 571–619 (hereinafter cited as *Fuel Investigation*). On October 1, 1944, ULR and ALT filed a Section 11(e) plan (Application no. 21) for the purpose of complying with the SEC Order of August 5, 1941. In this application (no. 21), ULR and ALT agreed to abide by the SEC Order, but they also proposed to organize a new pipeline company to transport natural gas from Hugoton field to Detroit. On June 26, 1947, ULR and ALT filed a new plan (Application no. 31) superseding the earlier plan. The new application proposed the continuance of ALT as a holding company over MichCon, Milwaukee Gas Light, Milwaukee Solvay, Mich–Wisc, and Austin Field Pipe Line Company. The SEC approved this application on December 30, 1947.

[3] See 19 SEC 366, 368–9 (1945); 73 PUR (NS) 324 (1947); 170 F. 2d. 453 (8th Cir. 1948); and 6 FPC 1, 37, 58, 74 (1947).

[4] 6 FPC 17.

[5] *Fuel Investigation*, pp. 642–4. In 1932, Woolfolk unsuccessfully attempted to acquire a 50 percent interest in Panhandle Eastern for American Light & Traction Company.

local and state political officials, that was not the case initially. The city attorney, William E. Dowling, foresaw in MichCon's pipeline proposal an attempt to recreate a gas monopoly for Detroit. "The whole plan," Dowling stated, "is one of absolute monopoly, controlling the life of the local utility from the source of supply through the pipe line to the local utility and the ratepayer's burner tips."[6] In reality, it was Panhandle Eastern which had enjoyed since 1936 monopoly power over natural gas deliveries to Detroit. But MichCon sought to recreate the cartel structure that had formally controlled midwestern gas supplies – hence the charge of "monopoly."

William Woolfolk, a former "trouble shooter" for Samuel Insull, and Maguire shared a mutual dislike of each other. Both were strong-willed men, who relished combat. Woolfolk now needed to prove that Panhandle Eastern's service was either unnecessary or insufficient so that he could receive federal certification to build a competing pipeline. Maguire, who was described by *Fortune* magazine as "a fighting, feuding, litigious Irishman," needed to prove the opposite if he wanted to protect this part of his market.[7] Maguire had already shored up his position by making Panhandle a profitable organization. It was the gas pipeline with the highest net profit per sales dollar, ranging from a high of $.23 in the early 1940s to $.16 in the late 1950s, consistently above average for long-distance pipeline firms. But efficiency alone could not protect his firm in a pervasively regulated market.[8]

The Detroit gas market and the fight over direct sales

Detroit was the location for numerous automobile and industrial plants which had contributed to the defense effort and were now an important part of the nation's burgeoning consumer society. Fears of a return to the Great Depression soon gave way in the postwar years to enthusiasm for economic growth and the widespread prosperity it was producing; it quickly became apparent that natural gas consumption in the

[6] SEC, "In the Matter of the United Light & Railways Company, American Light & Traction Company et al., November 13, 1947 (File nos. 59-11; 59-17; 54-25), tr 31-1832, 31-1850-1.

[7] "Natural Gas – Whoosh!," *Fortune*, December 1949, p. 109.

[8] "Natural Gas," *Forbes*, vol. 79, no. 1 (January 1, 1957), p. 63.

Detroit area would increase rather than remain stagnant or diminish. The postwar boom expanded the market as new homes were constructed and older homes were converted to natural gas.[9]

The original 1935 contract between Panhandle and MichCon had been the basis for the pipeline's first year of profitable operations: its net income increased from a negative $326 thousand in 1935 to a positive $1.7 million the following year. The fifteen-year contract originally specified that Panhandle would provide Detroit with a maximum delivery of 90 Mcf/d. This was increased in 1939 to 100 MMcf/d, and in June 1940 to 125 MMcf/d.[10] Between 1937 and 1945, MichCon thus had available more than 363 Bcf of natural gas but had accepted only 217 Bcf, about 60 percent of its allotment. Lacking facilities for storage, the utility firm only took delivery of the gas it could immediately resell.[11]

The utility's performance in this regard became a crucial element in Maguire's attempt to bypass MichCon and enter the Detroit market directly. His Detroit office – headed by his son-in-law, Vice-President Frederick H. Robinson – became the command headquarters for the public relations, political lobbying, and regulatory campaigns that this strategy required.[12] A

---

[9] See John H. Herbert, *Clean Cheap Heat: The Development of Residential Markets for Natural Gas in the United States* (New York: Praeger, 1992). Also see Walter J. McCarthy, *Detroit Edison Generates More Than Electricity* (New York: Newcomen Society in North America, 1983); and Robert J. S. Ross and Kent C. Trachte, *Global Capitalism: The New Leviathan* (Albany: State University Press of New York, 1990).

[10] *Fuel Investigation*, p. 620.

[11] Ibid. Also see 6 FPC 38.

[12] "The Fred Robinson Story" and "Personal History of Frederick Hampden Robinson," PECA. Panhandle Eastern established a Detroit office in 1942. Originally located in the United Artists building, the Detroit office had a permanent staff of only two persons. In 1946, Frederick H. Robinson, a company vice-president in charge of sales became the manager of the Detroit office. Robinson was not an ordinary employee, he was Maguire's son-in-law. In 1946, Robinson was named vice-president in charge of sales and soon thereafter assumed leadership of the Detroit office. Robinson remained in this position for four years. He was replaced by George Ballard, formerly Panhandle Eastern's sales manager, when Robinson transferred to the company's New York office. Born in southern California, he spent his youth in Nebraska and at his mother's home in Fredericksburg, Virginia. He at-

major point of contention was the maximum amount of gas available to the Detroit utility. Since MichCon only accepted full contract quantities in severely cold winter months, Panhandle was forced to operate its pipeline at considerably less than full capacity during the remaining months. The way out of this situation, as Maguire saw it, was for Panhandle to get up to full sales capacity by selling directly to Detroit-area industrial customers. In effect, he would skim off the cream, the largest customers, leaving MichCon to serve the rest of the market; years later, MCI would use this strategy and successfully break the Bell System's control of the national telecommunications market.

Maguire's reading of the Natural Gas Act indicated to him that direct industrial sales were legal and would fall under federal, not state or local, regulations.[13] The resulting conflict between Panhandle Eastern and MichCon was played out in FPC hearing rooms, court rooms, state public utility commissions, city councils, and the news media. The struggle involved fundamental issues of regulated competition and the protection of public–private monopoly; it arrayed an entrepreneurial firm against a well-ensconced utility and foreshadowed the battles that would accompany the deregulation movement of the 1970s and 1980s.

As early as 1940, Panhandle Eastern had started negotiations for selling surplus gas directly to Detroit Edison's Trenton Channel electric generating plant. This contract would have bypassed MichCon, and to Woolfolk, it represented an illegal and threatening "raid" on MichCon's territory. The nego-

---

tended Saint Johns College in South Africa then returned to the United States where he finished his education and entered the architectural landscape business. After this venture failed during the 1930s, he returned to his home in Fredericksburg, began working for the National Park Service, and began a courtship with Maguire's daughter.

[13] See Natural Gas Act (1938), 52 U.S. 821. Section 1 (b) of the act applies "to the sale in interstate commerce of natural gas for resale for the ultimate public consumption for domestic, commercial, industrial, or any other use. . . ." Sales made directly to industrial end users were not subject to FPC rate regulation but state and local commissions did regulate these sales. For an articulate industry view of federal rate regulation see Stuart M. Crocker, "Is Gas Rate Regulation Realistic?," *Public Utilities Fortnightly*, October 25, 1951, pp. 2–8.

tiations did not result in a contract, however, so the issue was for a time put on hold.

When Maguire renewed his assault, his position and that of his rival (and best customer!) were altered decisively by Mich-Con's effort to build its own supply line and squeeze Panhandle entirely out of the Detroit business. To do so, MichCon needed to prove that the existing gas supply was either inadequate or unnecessary if it was going to convince the FPC that a supply line was necessary. MichCon took the first course and attempted to show that Panhandle Eastern's gas supply was inadequate. This regulatory strategy forced MichCon to delay the development of a nearby underground gas storage facility at Austin Field. By storing off-peak gas purchases made in the summer, the utility would have had sufficient winter gas supply. With that goal in mind, MichCon had purchased (1941) the Austin Field in Mecosta County, Michigan, 140 miles from Detroit. But the utility put these reasonable plans on hold as it pressed for permission to build its own supply line and supplant Panhandle. In brief, an innovation that would have benefited Detroit consumers – especially during wartime – yielded to the politics of regulation.[14]

This struggle continued during the war, with Panhandle consistently opposing MichCon's efforts to acquire its own supply while repeatedly seeking to bypass the utility by signing contracts with Detroit industrial firms. The utility contended that after its contract with Panhandle expired (1951) and it completed its proposed Mich–Wisc line, it would not need gas supplied by Panhandle Eastern. In this dispute, local officials were wary of MichCon's gas supply strategy. Initially they blamed MichCon, not Panhandle, when wartime gas shortages plagued Detroit and surrounding areas. If the utility had built a line to the Austin Field and developed its gas storage capacities, they said, natural gas would have been available in emergency situations. As Dowling, the city's attorney said, "the very fact that the Panhandle Eastern contract allows 125 million cubic feet a day, and in some months in the summer they [Michigan Consolidated] only took 40 million cubic feet a

---

[14] Although the War Production Board (WPB) granted MichCon the steel priorities necessary to build a pipeline from Detroit to the Austin Field, the pipeline was not built due to MichCon's competitive strategy.

day, just about one-third of what they could have taken, displays [their attitude] very definitely."[15] Dowling believed that MichCon sought to create gas shortages which would justify the federal approval necessary to build the Mich–Wisc line and thereby integrate its Michigan and Wisconsin distribution systems. Panhandle made these charges look reasonable when it developed storage capacity in conjunction with another large Michigan utility, Consumers Power Company of Jackson, Michigan. Together, these two firms created the Michigan Gas Storage Company, which provided Jackson consumers with the same type of storage facility Detroit clearly needed.[16]

During the war, Panhandle was able to sell gas directly to two industrial firms in Michigan: Michigan Seamless Tube Company of South Lyon and Albion Malleable Iron Company. Both signed contracts after the War Production Board (WPB) requested the hookups due to the defense related production of the two firms. Since the federal government ordered these direct sales, and since the volumes were minimal, MichCon did

---

[15] Release to Foreign Language Press for week of September 3–October 5, 1946. Folder: "Inter-Racial Press of America," Box 189B10, PECA.

[16] Consumers, one of Panhandle's largest utility customers since 1939, desired increased storage capacity in order to compete more effectively with its Detroit-based rival, MichCon, and Maguire was only too happy to help. In early 1946, the two firms signed a contract resulting in the formation of the Michigan Gas Storage Company. The new company issued common stock in the amount of 8 million dollars, of which Panhandle bought 25 percent, or $2 million worth. The remaining 75 percent of Michigan Gas Storage was owned by Consumers. Panhandle Eastern was entitled to have two representatives on the Michigan Gas Storage Board of Directors. The Panhandle representatives were William Maguire and Ed Buddrus. Panhandle agreed to supply Michigan Gas Storage with 8.2 Bcf in 1946 and increase deliveries to 28.8 Bcf by 1950; in 1959 annual deliveries to the storage facility reached 80 Bcf per year.

For background on Panhandle's other storage facilities, see "Panhandle Eastern to Supply Gas to Michigan Gas Storage Company," *Panhandle Lines*, February 1947, p. 3; *Panhandle Lines*, August 1946, p. 8; Stevenson, Mr. Interview by Robert B. Eckles, February 7, 1956, PECA; "Natural Gas for Tomorrow," *Panhandle Lines*, October 1961, pp. 4–5; and "1960 Headlines in Review," *Panhandle Lines*, December 1960. For background on Consumers Power, see George Bush, *Future Builders: The Story of Michigan's Consumers Power Company* (New York: McGraw–Hill Book Company, 1973).

not object. It did, however, oppose Maguire's repeated efforts to make direct delivery of natural gas in Detroit. Maguire wrote to Detroit Mayor Edward Jeffries explaining that to assure Detroit's residential customers of the lowest possible prices, Panhandle Eastern should operate at full capacity in the summer. "This can be done through the direct sale of gas on an interruptible basis to large industrial customers in and around Detroit." Maguire offered to pay the city a fee for the privilege of laying and operating pipelines under Detroit streets and alleys. His firm, he said, could provide fuel "at very favorable rates compared with those prevailing at Detroit now."[17]

MichCon responded by stating publicly that it did "not contemplate renewing the Panhandle contract." This statement evoked intense criticism from both the Detroit Common Council and Michigan congressmen. Congressman John Lesinski charged that

> For approximately 15 years the Michigan Consolidated Gas Co. at Detroit, and its predecessor, Detroit City Gas Co., has been conducting a ruthless form of guerrilla war against every threat of competition for the great gas market which it dominates, a deliberate campaign in which it has had the assistance of the powerful Columbia Gas & Electric Corporation, and, more recently, of others whose activities merit attention.[18]

On October 20, 1945, Maguire appeared to score a major victory. Panhandle signed a 25 MMcf/d contract with Ford Motor Company's Dearborn plant, and early in 1946, the firm filed an application for an FPC certificate authorizing the construction of a connection between Panhandle's River Rouge meter station and Ford's River Rouge plant – a line which would run parallel to MichCon's own connection to Ford.[19] Panhandle said it intended to begin gas deliveries by May 1, 1946.

MichCon counterpunched by observing that this intrusion

[17] W. G. Maguire to Edward J. Jeffries (mayor of Detroit), January 4, 1945. Folder "PE–MC Data, Prior to 1948" Box 189B10, PECA. The letter is also quoted in Detroit Common Council, *Minutes*, February 21–2, 1944, Box 189B5, PECA. In 1945, Panhandle Eastern's sales to Detroit represented 27 percent of its total sales.

[18] Representative John Lesinski (MI) lambasted MichCon for monopolistic practices. See *Congressional Record*, House, 79th Cong., 1st sess., vol. 91, no. 26, (February 12, 1945), pp. 1056–60.

[19] 5 FPC 45.

on its market was an effort to undermine the integrity of the Michigan Public Service Commission (MPSC). The utility filed a formal protest with the FPC and the MPSC on December 18, 1945, and a few days later the MPSC asked Panhandle to show why a "cease and desist" order should not be issued. At the FPC's public hearings (January 7–12, 1946) several intervenors protested the proposed gas sale. The same dialogue took place at the state level as well. Panhandle argued that it was a contractor, not a public utility, and that the MPSC had no jurisdiction over an interstate pipeline. But this was a distinction one could hardly expect a regulatory agency to accept. On February 18, the MPSC ordered Panhandle to stop contracting direct sales to Michigan's industries "unless a [state] certificate of public convenience was issued permitting the performance of such service."[20] The pipeline had no choice but to withdraw from the contract with Ford.[21]

As Maguire saw it, he had lost a battle, but not the war. He persisted and was able to complete a twenty-year contract with a utility company located in Chatham, Canada. This time the FPC granted a certificate to export to Union Gas a maximum of 5.5 Bcf yearly, provided the needs of present or future customers in the United States were met first. Although Union Gas intended to resell a portion of this gas, most was dedicated to Union's "Dawn Field" underground storage facility for subsequent retrieval during peak-load requirements.[22] The City of

[20] "Opinion and Order to Cease and Desist," State of Michigan, Michigan Public Service Commission, February 18, 1946, Folder: *PEPL Co. v. Michigan Public Service Commission*, Box 87B7.

[21] 341 U.S. 329. In Indiana and Illinois, direct sales were curtailed or modified by state commissions. The Illinois state commission prevented Panhandle Eastern from selling direct to a company in Decatur on the principle that such a sale ought to be made through the domestic distributor. In Indiana, the state supreme court ordered Panhandle Eastern to get state commission approval to sell gas interstate. Generally, the Indiana and Illinois state commissions sought to protect the local distributor.

[22] "Panhandle to Export Gas to Canada," *Panhandle Lines*, August 1946, p. 3; and "Detroit River Crossing," *Panhandle Lines*, January 1947, p. 5. Panhandle constructed a sixteen-inch pipeline from its Detroit Regulator Station to River Rouge, Michigan. There, two parallel pipelines extended under the Detroit River to the Canadian shore at Ojibwa, Ontario to a pipeline connection with Union Gas Company's system in Windsor, Canada. Panhandle contracted to deliver the gas during the off-peak summer months.

Detroit – but not MichCon – had objected to this contract as well, on the grounds that it would lessen the amount of natural gas Panhandle Eastern could make available to MichCon.[23] But the FPC was unwilling to block the sale on these grounds alone.

Direct industrial sales that bypassed the utility were another matter, and the FPC refused to undermine the regulatory system even though Panhandle would have been able to lower the price of gas to Detroit's industrial customers. It would possibly have been able to ensure more reliable supplies. A fuel crisis erupted in Michigan during the winter of 1947, when a number of manufacturing plants were temporarily forced to close. By February 1948, ninety-five industrial plants in the city were shut down wholly or partially because of the natural gas shortage. This time both the Detroit Common Council and the Detroit media accused Panhandle of creating the shortage. The firm countered with an advertising campaign in local newspapers charging that MichCon, not Panhandle Eastern, deserved blame for the shortage. William E. Dowling, Detroit's Corporation Counsel, sided with Panhandle and asserted that the "present shortage of gas is of artificial creation." The "onus" he said, was "upon the local company [Michigan Consolidated]."[24] Nevertheless, the bad publicity did not help Panhandle's cause at a time when it was once again trying to sign industrial contracts.

Although Maguire had been defeated on this issue before, he forced the federal and state regulators to return to the problem again. Panhandle signed a contract with Great Lakes Steel

---

[23] *Fuel Investigation,* pt. I, p. 591. MichCon did not oppose the contract. MichCon attorney Harry A. Montgomery later told a congressional panel in response to a query regarding the contrast of MichCon's silence to the Detroit City Council's opposition, that "We had pending before the Federal Power Commission at that time an application for a certificate of public convenience and necessity to build [a] new pipe line that Panhandle had forced us to build.

So, in applying for a certificate to build our own line we had to show the Federal Power Commission that we had the gas, and that we had the markets for it, and that our pipe line could supply those markets. Now, if at the same time we had gone in before the Federal Power Commission and objected to the sale of gas to Canada we would be taking two contradictory positions at the same time in the same forum."

[24] Release to Foreign Language Press for week of September 3–October 5, 1946. Folder: "Inter-Racial Press of America," Box 189B10, PECA. See *Detroit Times,* May 31, 1952.

Company. MichCon opposed this deal of course.[25] So did the FPC, which warned Panhandle to secure regulatory permission before continuing with this plan. As the secretary of the FPC noted, Panhandle would have to prove public convenience to sustain the contract and to prove necessity before it could begin construction of the necessary facilities. Although the Commission subsequently approved an application by Panhandle to sell gas to a new Corning Glass Plant in Albion, Michigan, it refused to budge where the Detroit market was concerned.[26]

Maguire signed a second contract (announced on June 9, 1950) for direct delivery of surplus gas to Ford's River Rouge plant. Once more the familiar contestants mounted their now familiar arguments. Once more Panhandle was rebuffed.[27] This time the bad news came from higher authority; in the spring of 1951, the U.S. Supreme Court ruled that Panhandle Eastern must receive a state certificate of public convenience and necessity for making direct industrial sales. "It does not follow," wrote Justice Minton, "that because appellant is engaged in interstate commerce it is free from state regulation or to manage essentially local aspects of its business as it pleases."[28] Once again, the MPSC refused to grant Panhandle the permission it needed to proceed with the sale. Maguire – who had gained control of the pipeline because of the new political environment created by the New Deal – was at last thwarted by that same political system. This market had been politicized and Maguire could not even cut off service to MichCon without FPC approval – a ruling he was unlikely to receive.

Nor could he thwart the entrepreneurial plans of his rival by way of regulatory politics. On February 19, 1945, American Light & Traction had filed with the FPC an application for a certificate of public convenience and necessity to construct its Michigan–Wisconsin pipeline from the Hugoton field near

[25] "Two Firms Renew Fight over Gas," *Detroit Sunday Times*, February 5, 1950, pt. I, p. 6; and "Panhandle Seeking Deal with Great Lakes," *Detroit Free Press*, February 10, 1950.

[26] "FPC OK assures New Albion Plant," *Detroit Times*, March 10, 1950. Panhandle contracted with Corning for 6 MMcf/d at $.32 per Mcf.

[27] *Detroit Free Press*, September 2, 1951; and 44 N.W. 2d. 324.

[28] U.S. Supreme Court, no. 486, October term, *Panhandle Eastern Pipe Line Co.* v. *Michigan Public Service Commission*, May 14, 1951. Also see "State Upheld on Gas Sales," *The Detroit News*, October 11, 1950.

Guymon, Oklahoma to points near Toledo, Ohio, and Detroit. From Detroit, the proposed line would continue to the Austin Storage field near Big Rapids, Michigan, and in April of 1945 the application was amended to include service to communities in Missouri, Iowa, and Wisconsin. At that time, no interstate pipeline sold gas to Wisconsin customers. Asked during SEC hearings why Wisconsin had no natural gas, Woolfolk explained, "If the majority of the Board were coke oven inclined people, until they died, of course natural gas influence could not be successful."[29] Woolfolk also renewed his promise – expressed to the Detroit press – that MichCon had no intention of renewing its contract with Panhandle Eastern.

Panhandle fought a bitter campaign against the new pipeline. It contended that the FPC had no authority to grant such a certificate.[30] Mich–Wisc management meanwhile signed a contract to supply MichCon's total gas requirements after December 31, 1951. It was now clear that MichCon might shortly be able to achieve its goal of pushing Panhandle out of the Detroit market entirely.[31] With his back against the wall, Maguire continued the struggle. After the completion of the hearings, Panhandle filed a formal motion for dismissal. The company based its motion on three propositions: First, that the FPC could not deprive Panhandle of its "grandfather" rights to the Detroit market. Second, Panhandle was able and willing to perform the

[29] 8 FPC 1–91. SEC, October 9, 1947, TR 31–2027. Also see "State Loses Point on Gas," *Milwaukee Journal*, December 13, 1951. The Michigan–Wisconsin line would serve utility customers in Wisconsin (8), Michigan (3), Iowa (4), and Missouri (1). The state of Wisconsin also discouraged natural gas use through the imposition of a tax. In 1943, the state government set a $.07 per Mcf tax on natural gas, presumably at the behest of coal and coke interests in the state. This tax was repealed in 1947. See Wisconsin Laws of 1943, c. 339; Wisc. stat. 76.55 (1943); and 60 PUR 291.

[30] For a series of newspaper articles which chronicle the MichCon–Panhandle Eastern conflict see *Detroit News*, February 21, 1945; April 3, 1945; June 22, 1945; and September 20–1, 1945; March 4, 1947; *Detroit Free Press*, February 22, 23, 24, 1945; April 3, 1945; September 24, 1945; May 16, 1946; September 24, 1946; *Detroit Times*, April 2, 1945; May 16, 1946; and September 24, 1946.

[31] *Fuel Investigation*, p. 477. For an informative discussion of competition in the natural gas industry, see John T. Miller, Jr., "Competition in Regulated Industries: Interstate Pipelines," *Georgetown Law Journal*, vol. 48, no. 2 (Winter 1958), p. 224.

service proposed. Third, that Michigan–Wisconsin was not able to perform the service included in its application.[32] NGPL and Northern Natural filed similar motions.

On November 20, 1946, the FPC decided in favor of economic growth, against monopoly, and against grandfather rights to a market. Maguire had lost again. Commissioners Leland Olds and Claude Draper dissented and found in favor of Panhandle Eastern. But the three other commissioners voted for the Mich–Wisc line.[33] On December 30, 1947, the SEC, reversing an earlier ruling, approved American Light & Traction's plan to continue operating the Milwaukee Gas Light Company along with MichCon and Mich–Wisc.[34] In January of the following year Panhandle made another major overture to renew the pipeline's contract with MichCon, offering to sell gas at $.185 per Mcf (compared to Mich–Wisc's price of $.21). There was no response.[35] But, paradoxically, in 1949 the utility decided it wanted to continue Panhandle's gas deliveries after the completion of the Mich–Wisc line.[36] Dramatic increases in demand prompted this change in strategy, a change that still left Maguire unhappy. Panhandle belatedly gave notice that it intended to reduce deliveries to MichCon from 127 MMcf/d to 87.5 MMcf/d, but the FPC demurred. The Commission forced the pipeline to continue abiding by the terms of its original – now expired – contract due to MichCon's increased need for gas to supply its residential, commercial, and industrial customers.[37] Maguire was boxed in, able neither to penetrate the market that interested him nor to leave the utility customer with whom he had been fighting for many years. There were glimmerings in these events of the manner in which the regulatory system seemed likely to evolve in the years ahead, and some of these glimmerings were threatening to entrepreneurs like Maguire.

[32] Federal Trade Commission Reports, pp. 24–5.
[33] 6 FPC 37–40, 40–50, 74–91. See FPC Order, November 30, 1946.
[34] 10 SEC 1138; 12 SEC 218.
[35] See 6 FPC 1–7 and 10 FPC 313. For a number of related newspaper stories see *Detroit News*, August 2 and 4, September 26, and October 21, 1947; *Detroit Times*, September 13 and 25, 1947; and *Detroit Free Press*, August 26 and October 22, 1947.
[36] American Light & Traction became American Natural Gas on June 15, 1949. 9 SEC 833, 856 (1941).
[37] 11 FPC 172 and 6 FPC 91. Also see 236 F. 2d 289 and 335 U.S. 854.

World War II and the transformation of the industry

Not all of the signs were threatening. The entire natural gas industry had been reshaped by the wartime experience, and Panhandle was well positioned to take advantage of these developments. During the war, natural gas production had increased by 55 percent.[38] The demand was especially strong in the highly industrialized Appalachian region – sometimes referred to as the American Ruhr – which on the eve of the war consumed approximately 18 percent of U.S. production. By that time, however, the region's shallow gas-bearing formations were largely depleted. Natural gas produced from the region's deeper formation, the Oriskany, peaked in 1942 and fell rapidly in 1943.[39]

Shortages were a problem in this highly industrialized area. The states of Ohio, New York, Pennsylvania, West Virginia, Maryland, Virginia, and Kentucky contained 660 war factories. Approximately 184 steel mills in Pittsburgh, Youngstown, and Wheeling used gas as a fuel. So too did chemical and metallurgical works located in western New York and rubber factories in Akron, Cincinnati, Cleveland, Dayton, and Toledo. These factories and other consumers in Appalachia were using as much as 400 Bcf per year, and as war production got under way, new sources needed for Appalachian industry.[40]

Primarily responsible for solving their problem was Secretary of the Interior Harold I. Ickes, who was coordinating oversight of the wartime energy industry through the Petroleum Administration for War (PAW). Ickes had persuaded President

[38] John W. Frey and H. Chandler Ide, *A History of the Petroleum Administration for War, 1941–1945* (Washington, D.C.: U.S. GPO, 1946), p. 227.

[39] Ibid., pp. 228–9.

[40] For a general discussion of the federal government's special war time energy regulatory measures see Frey and Ide, *A History of the Petroleum Administration for War, 1941–1945*. For a collection of documents related to government war planning, see Natural Gas Industry Advisory Committee, Summaries, May 11, 1943 and August 3, 1943, Folder: "623.5 – Natural Gas: Industry Advisory Committee Meetings," Box 2091, RG 179, NARA. For a description of the War Production Board's evolving administrative structure for natural gas, see Folder: "037.68, Natural Gas Division, War Production Board, Reports and Issuances," Box 247, RG 179, NARA. For relations between the FPC and WPB see W. L. Batt to Leland Olds, January 2, 1942, Folder: "760 – Federal Power Commission," Box 2277, RG 179, NARA.

Roosevelt to create such an agency, and F.D.R. subsequently chose him to be the first head of the organization. Ickes had special powers over virtually every aspect of the oil and gas industries and deserved his unofficial title of oil "czar." While many businessmen were apprehensive about Ickes's antibusiness reputation, many were pleased when he chose Ralph K. Davies of Standard Oil of California to be his deputy administrator.[41]

PAW shared power over natural gas with the Supply Priorities and Allocation Board (SPAB). The SPAB, which later merged into the War Production Board (WPB), had broad authority over industrial production, including allocations of steel to companies seeking to build pipelines. The agency dictated specific gas sales allocation orders to natural gas lines and concern over energy shortages prompted the WPB to issue conservation orders. On February 16, 1942, Order L-31 requested that utilities voluntarily make pooling arrangements "to achieve practicable maximum output in the area or areas in which a shortage exists or is imminent."[42] Under this order the WPB could integrate natural gas systems, curtail sales when necessary, and reallocate existing sales.[43]

As PAW and WPB attempted to deal with the Appalachian problem, the agencies tried several approaches; one was to increase output. Between 1942 and 1945, approximately 70 percent of all gas wells drilled in the country were in Appalachia. Increased drilling activity, however, did not significantly raise production levels, and the PAW was forced to issue conservation guidelines. These too failed to solve the regional supply problem.[44]

---

[41] Harold L. Ickes, *Fightin' Oil* (New York: Alfred A. Knopf, 1943).

[42] "Limitation Order L-31," Folder: "Natural Gas – L-31," Box 6, RG 179, NARA.

[43] Ernest R. Acker, "The Gas Industry in War," *Proceedings of the American Gas Association* (1942). In the summer of 1942, the WPB issued a similar order, L-174, which was modified after L-31 but imposed the same restrictions on the manufactured gas industry. Essentially, both L-31 and L-174 greatly limited new gas sales for nonmilitary or nonmobilization purposes. See J. A. Krug to J. S. Knowlton, no date, Folder: "Natural Gas – L-31," Box no. 6, "L," RG 179, NARA; and Memorandum by General H. K. Rutherford, January 13, 1942, Folder: "Limitation Order L-31," RG 179, NARA.

[44] Frey and Ide, *A History of the Petroleum Administration for War, 1941–1945*, pp. 229–30.

A third strategy involved using pipelines to bring natural gas from the abundant southwestern gas fields to Appalachia, and one of these early ventures succeeded. This project started in April 1940, when Curtis Dall, recently divorced from Franklin D. Roosevelt's daughter, Anna, established the Tennessee Gas Transmission Company. Dall spent more than three years in an unsuccessful attempt to acquire a certificate of public convenience and necessity from the Federal Power Commission. Dall's efforts were both helped and hindered by the war. For a time, he could not get the War Production Board (WPB) to allocate steel to his pipeline and thus could not arrange the conditional financing or the gas purchase and sales contracts he needed to persuade the FPC to issue a certificate.[45]

While Dall and his associates struggled to promote their new line, the WPB was looking for less expensive and less time-consuming ways to get gas to the Appalachian region.[46] The agency requested that either Ohio Fuel Gas Company, a subsidiary of Columbia Gas, or Panhandle Eastern construct a line to interconnect their systems at a point near Toledo, Ohio.[47] The WPB issued an order providing for the necessary materials before the FPC conducted hearings for the project. In fact, the project was considered so vital that the pipe was being rolled even before the hearings began.

Although the WPB set the parameters of the project, only the FPC could certificate a firm to build and operate the line. The FPC held hearings at which representatives from both Panhandle Eastern and Ohio Fuel presented proposals. Pan-

[45] Castaneda, *Regulated Enterprise*, pp. 34–65.

[46] For a collection of correspondence relating to a variety of efforts, including those of Tennessee Gas, to transport fuel from the southwest to the northeast during the early years of the war, see Folder: "TG&TCO. War Production Board No. 55,444," Box 2919, RG 253, NARA. Also see "Summary of the Findings, Policies, and Recommendations of the Fuel and Utilities Branch of the Office of Civilian Supply," Folder: "Public Services Division, WPB-Functions," Box 202, RG 179, NARA.

[47] In one respect, the WPB's request was ironic. At the time, both Panhandle Eastern and Ohio Fuel Gas Company remained subsidiaries of the Columbia Gas & Electric Corporation. Although Columbia Gas would shortly sell all of its Panhandle Eastern stock, the dispute involving Panhandle Eastern, Columbia, Mo–Kan, and Michigan Gas was not allowed to hinder the wartime cooperation which the WPB demanded from the industry.

handle estimated the job's cost at \$394,000; Ohio Fuel, at \$410,815. The Ohio Fuel representatives requested that the FPC consider the project only a wartime emergency measure, while Panhandle's officials told the Commission that the extension should be permanent and authorized by a regular certificate. Appalachian reserves would continue to deplete even after the war, Panhandle contended.[48]

Panhandle won this round of regulatory politics. The FPC granted the firm a conditional certificate, forcing the physical connection of the Panhandle and Columbia Gas systems. Columbia had long resisted this innovation, and Maguire had long desired this opportunity to penetrate new markets. The FPC certificate was only for five years, but as soon as the connecting line was built, gas could flow. The FPC approved the certificate on October 2, 1942. On February 4, 1943, Panhandle and the East Ohio Gas Company (formerly of Standard Oil) signed a contract providing for deliveries of 50 MMcf/d.[49]

This, however, was only a stopgap solution to the Appalachian fuel crisis. "It is crystal clear," the FPC stated, "that additional natural gas is needed in the Appalachian region. It follows, therefore, that a realistic view of this situation definitely shows that the public convenience and necessity will be served by the construction and operation of . . . [Tennessee Gas's] pipeline into the area . . ."[50] Unable to raise the money

---

[48] 4 FPC 303.

[49] 4 FPC 301–9. The East Ohio Gas Company was an affiliate of Consolidated Natural Gas (formerly Standard Oil of New Jersey's natural gas operations), and Ohio Fuel was affiliated with Columbia Gas & Electric. The regulatory agency required that the pipeline be constructed so gas could flow either way and not interfere with either Panhandle Eastern's existing customers or state and local authority over other natural gas operations. Intervenors from the coal and railroad industries did not oppose the pipeline connection at the hearings, but the National Coal Association pleaded that the certificate not be extended beyond the war emergency. The FPC, however, later dropped the five year limitation on the certificate.

[50] 4 FPC 446. Also see Frey and Ide, *A History of the Petroleum Administration for War*, pp. 231–2. For some insight into the WPB's oversight of the wartime gas shortage see the following correspondence: J. A. Krug, director, Office of War Utilities, to Paul R. Taylor, Director, Natural Gas Division, Office of War Utilities, July 10, August 9, August 23, September 7, 1943 and January 7, 1944, Folder: "037.68, Natural Gas Division – Bi-Weekly Reports, July 10, 1943 – December 31, 1943," Box 247, RG 179, NARA.

needed to fund its project, the original Tennessee investors sold their undertaking to the Chicago Corporation, which quickly moved ahead with the project. Construction of the Tennessee Gas line began on October 31, 1944, and little more than one year later, the company began delivering fuel to Appalachian area distribution companies.[51]

The gas deliveries of Panhandle Eastern and Tennessee Gas into Appalachia helped alleviate the regional natural gas shortage, but demand for the fuel continued to increase at a rapid pace. During the severe winter of 1944–5, there was another crisis. In response to the shortage on the Columbia System, the WPB directed Panhandle to curtail its interruptible industrial customers. More trouble followed. On January 4, Columbia's line supplying Pittsburgh blew out; near disaster struck again on February 2, as MichCon was forced to cut off gas to its industrial service area and redirect it into the Appalachian region to prevent severely low system pressures. The Governor of Ohio closed all state offices for three days in order to conserve natural gas.

Concerned about these serious problems, the WPB early in 1945 asked Panhandle to expand its pipeline facilities and transport more natural gas into Appalachia. The agency wanted the firm to add additional compressors and looped lines for "ameliorating the shortage [of natural gas] which otherwise will exist in the Appalachian area next winter." The government wanted the firm's plan at the "earliest possible moment."[52] Maguire was happy to do so. On February 2, the firm proposed a comprehensive expansion of its system – an expansion that involved a new compressor as well as pipelines.[53]

---

[51] Frank H. Love, "Construction Features of the Tennessee Gas and Transmission Company Pipe Line," *The Petroleum Engineer*, vol. 16, no. 2 (November, 1944), pp. 121–44. Also see Tenneco, Inc., "Tenneco's First 35 Years," 1978.

[52] 4 FPC 268.

[53] 4 FPC 264–5. The FPC opened hearings on Panhandle's application on February 15 and the hearings ended on February 23. Since Panhandle's expansion plans were based on a WPB request, the WPB granted Panhandle Eastern a high preference rating soon after it submitted its plans. The original connection was located near Maumee, Ohio and the other was at the Ohio–Indiana state line near Muncie, Indiana. Based on a contract dated February 9, 1945, Panhandle Eastern would deliver 25 MMcf/d to Ohio Fuel for twenty years and an additional 25 MMcf/d for five years.

Gulf Coast reserves and Trunkline Gas Company

Nor did Maguire's plans for the postwar economy stop there. In 1946 he moved Panhandle's executive offices from Chicago to New York City, where it would be easier to obtain the financial support he needed to make the company a major player in the new effort to link eastern markets with burgeoning supplies of Gulf Coast gas (Table 5.1). Thwarted in Detroit, he would look elsewhere for opportunities to expand. The Tennessee Gas Transmission Company, built during the war, had shown the way; now Panhandle and others would follow.

Two firms did rush to develop this business: Texas Eastern Transmission Corporation and Transcontinental Gas Pipeline Company. Texas Eastern bid for and won two government-financed, war emergency pipelines originally constructed to tap into Gulf Coast oil production. The Big Inch was a twenty-four-inch diameter crude oil line extending 1,254 miles from Longview, Texas, to Linden, New Jersey. The Little Big Inch was a twenty-inch refined-products-line connecting the Beaumont area on the Texas Gulf Coast with Linden, New Jersey.[54]

Texas Eastern Transmission Corporation – headed by Herman and George Brown of the Houston-based construction firm Brown & Root – presented the high bid of $143,127,000, which was approximately the same amount the government had paid for the construction of the lines. After Texas Eastern won the bid, Transcontinental built a third pipeline extending from the Texas Gulf Coast into the New York area. These two firms and Tennessee Gas then embarked on a vigorous competition for the vast northeastern gas markets in Pennsylvania, New York, and New England.[55]

Briefly, Texas Eastern and Panhandle entered into an emergency exchange agreement (in 1947), but Maguire already had in mind tapping Gulf Coast natural gas with his own new pipeline.[56] He soon discovered an opportunity to do so in a speculative venture developed by a small group of independent businessmen led by Ralph Davies, former Deputy Administra-

---

[54] Castaneda and Pratt, *From Texas to the East*, pp. 17–48. Also see Arthur M. Johnson, *Petroleum Pipelines and Public Policy, 1906–1959* (Cambridge, Mass.: Harvard University Press, 1967), pp. 322–7.

[55] Castaneda, *Regulated Enterprise*, pp. 66–93.

[56] 6 FPC 119–21.

Table 5.1. *Marketed production of natural gas in the Southwest and Northeast, 1935–1953 (Bcf)*

Year	Northeast	Southwest	U.S.
1935	276	1,172	1,917
1936	318	1,314	2,168
1937	341	1,476	2,408
1938	298	1,441	2,296
1939	331	1,535	2,477
1940	354	1,679	2,660
1941	378	1,743	2,813
1942	394	1,907	3,053
1943	415	2,151	3,415
1944	372	2,418	3,711
1945	342	2,658	3,919
1946	377	2,727	4,031
1947	394	3,044	4,582
1948	391	3,511	5,148
1949	344	3,805	5,420
1950	353	4,489	6,282
1951	385	5,419	7,457
1952	344	5,981	8,013
1953	350	6,318	8,397

Source: American Gas Association, *Historical Statistics of the Gas Industry* (New York: American Gas Association, 1961), pp. 52–3.

tor of the PAW and vice-president of Standard Oil of California (SOCAL). Davies, along with Lewis MacNaughton and Everett DeGolyer of the Dallas-based oil and gas consulting firm of DeGolyer and MacNaughton; P. McDonald Biddison, a leading pipeline engineer; and Carl I. Wheat, a Washington based oil and gas attorney, had together purchased a collection of large South Texas gas reserves. The five men wanted to make an agreement with an existing pipeline firm to help them finance the construction of a new line that would purchase all of their Gulf Coast gas.[57] They incorporated the Trunkline Gas Supply Company in Delaware on March 12, 1947, and elected Davies as president. They quickly filed for a certificate of public convenience and necessity with the FPC.

[57] Eckles, "Subsidiaries and Affiliated Companies," p. 3, IHS. Also see the Corpus Christi *Caller–Times*, October 14, 1956, p. 7E.

Figure 5.1 Boring under a highway during Trunkline construction, 1951

For the next two years, the Trunkline project floundered.[58] It was near collapse in the fall of 1949, when Maguire learned about its desperate need of financing and guaranteed markets. He acted quickly. On October 31, 1949, he and Davies agreed that Panhandle Eastern would purchase 60 percent of Trunkline's outstanding stock. Fifteen days later, the remaining 40 percent of Trunkline's stock was placed in a voting trust consisting of Maguire, Byrd, and Davies. The final result was Panhandle's ownership of 96 percent of Trunkline's common and preferred stock. The remainder was owned by Standard Oil of California, a tribute to Davies's intervention and influence. The FPC now issued a certificate for Trunkline, setting Panhandle's compressor station at Tuscola, Illinois as its new terminus.[59]

[58] For sources which trace the early travails of the Trunkline project see "New Company Headed By Davies Plans Wholesale Gas Line," *Oil and Gas Journal*, March 29, 1947, pp. 184–5; 9 FPC 105–26; and 8 FPC 260–1.

[59] Trunkline, *Minutes*, November 1 and November 15, 1949, and Panhandle Eastern, *Minutes*, November 22, 1949. "Petition to Amend Order Issuing

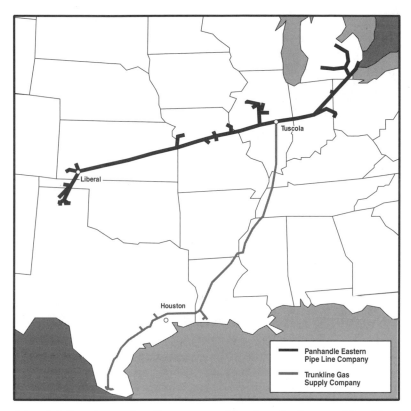

Map 5.2 Trunkline Gas Company and Panhandle Eastern Pipe Line
Company, 1950

Panhandle announced the acquisition on February 27, 1950
(Map 5.2). At the company's annual meeting in March, the stockholders learned that their company had arranged a $12,500,000
bank loan to finance the purchase, and that the estimated cost
of the 740-mile, twenty to twenty-six-inch Trunkline system

Certificate of Public Convenience and Necessity," Federal Power Commission, Docket no. G-882, p. 3, Box 189B8, PECA. On July 5, 1950 Trunkline's
official name became Trunkline Gas. Carol M. Loeb–Rhoades & Co. and
Metropolitan Life Insurance Co. of New York agreed to hold most of the stock
on February 2, 1950. The FPC issued the new certificate on May 4, 1950. See
Milt C. Allen, interview by Robert B. Eckles, November 3, 1956, PECA.

Figure 5.2 William G. Maguire turns the gas flow valve at Trunkline dedication, 1951

would be $85,000,000.[60] Financing had been accomplished by contracting with a group of insurance companies headed by Metropolitan Life Insurance. The insurance group had subscribed to a total of $61,000,000 in Trunkline's first mortgage bonds.[61]

On October 1, 1951, in Tuscola, Illinois, William Maguire opened a valve which initiated Trunkline's transmission of Gulf Coast gas into the Panhandle Eastern pipeline system. Now Panhandle, like the other major pipeline firms, would be able to take advantage of the growing demand for natural gas and the enormous supplies available in the Gulf Coast region of

[60] Panhandle Eastern, "Press Release," (Detroit office, 12 P.M.) February 27, 1950, Box 189B8, PECA.
[61] For an analysis of life insurance company financing of gas pipelines after World War II, see Richard W. Hooley, *Financing the Natural Gas Industry: The Role of Life Insurance Investment Policies* (New York: Columbia University Press, 1961).

Table 5.2. *Growth of the transmission system, 1952–1968*

	Panhandle Eastern	Trunkline	Industry
1952 total sales (both lines)[a]	321		———
transmission line (miles)	3,100[b]	1,300	368,900
compressor horsepower	271,460	32,000	n/a
1954 total sales (both lines)	319		———
transmission line (miles)	3,290[b]	1,300	420,900
compressor horsepower	277,100	38,350	4,255,000
1956 total sales (both lines)	400		———
transmission line (miles)	6,135	1,358	490,500
compressor horsepower	312,000	95,000	4,838,000
1958 total sales (both lines)	412		———
transmission line (miles)	6,300	1,400	540,800
compressor horsepower	315,000	95,000	5,612,000
1960 total sales (both lines)	410		———
transmission line (miles)	6,600	2,300	608,000
compressor horsepower	361,000	98,000	6,359,000
1962 total sales (both lines)	578		———
transmission line (miles)	7,500	2,700	663,600
compressor horsepower	388,000	126,000	7,064,000
1964 total sales (both lines)	701	[c]	———
transmission line (miles)	10,800	[c]	724,500
compressor horsepower	600,000	[c]	7,546,000
1966 total sales (both lines)	829	[c]	———
transmission line (miles)	11,900	[c]	789,200
compressor horsepower	750,000	[c]	8,182,000
1968 total sales (both lines)	973	[c]	———
transmission line (miles)	13,000	[c]	852,400
compressor horsepower	815,000	[c]	9,146,000

[a]Total sales figures in BCF; Source: Various Company Annual Reports.
[b]These mileage figures include only main transmission line.
[c]Data for Trunkline in 1964, 1966, and 1968 is shown as part of data for Panhandle Eastern in those years.

the United States. Since the late 1940s, the company had increased its drilling and exploration activities. By 1951, it owned significant proved gas reserves and would acquire more in the years ahead.[62] The firm had meanwhile made substantial technological improvements in its transmission system,

[62] T. M. Thawley, "A History of Panhandle Eastern Pipe Line Company" (company document), January 1959, p. 22, PECA.

improving its communications with a radio-telephone network and pioneering a new gas flow formula.[63]

Judging on the basis of Panhandle's development, the new role of the government had to that point dramatically altered but certainly not eliminated entrepreneurship in this industry. The expansion of the national economy, the increased demand for cheap energy, and the availability of enormous supplies of southwestern gas had continued to create unusual opportunities for innovations. During the war, a significant amount of the innovation had been financed and directed by the federal government. What the new regulatory system had changed was the manner in which innovation took place, and the firms and individuals which most benefited from change. Regulatory politics, not market forces, determined that MichCon would be able to acquire a pipeline and that Panhandle would not be able to sign contracts with industrial customers in Detroit. On the other hand, wartime controls enabled Panhandle to push into the Appalachian market following the war, and the FPC had not blocked Maguire's efforts to tap Gulf Coast supplies and expand the Panhandle system. As late as 1951, then, the new structure of regulatory controls was gradually reshaping the industry, without blocking the powerful forces of growth. In that context, Maguire and Panhandle had lost several battles but had become seasoned political as well as economic warriors. The firm and its executives were well-positioned to become major players if the industry's expansion continued.

[63] See Eckles, Pt. II, C. VI, p. 7, IHS. Also see American Gas Association, *Transmission: Pipelines/Planning and Economics*, vol. 2, Book T-1 (Arlington: American Gas Association, 1989), p. 17. Also see "Panhandle's Flow Formula Used by Tennessee Lines," *Panhandle Lines*, December 1944, p. 4. Known as the "Panhandle Formula," it was developed over a period of years by a number of Panhandle Eastern employees. In the late 1930s, Thomas Weymouth devised the original flow formula. In 1940, Panhandle's C. H. M. Burnham, chief engineer, and F. S. Young, assistant chief engineer, developed the so-called Panhandle Flow Formula based upon the work of Japanese engineer S. Nikuradse. In 1955, Panhandle's John F. Schomaker and L. E. Hanna made improvements (for larger diameter pipe) based on internal test data from the company's pipelines and published the modified Panhandle formula.

6

Regulatory dynamics and market control

During the "American Era," through the late 1960s, it appeared to most Americans that this industry and most others could continue to grow forever. The entire U.S. economy grew dramatically. The Gross National Product (GNP) rose from $227 billion in 1940 to $726 billion in 1969. Government spending for defense, highway building, and the space program helped fuel that growth, as did consumers, anxious to partake of the benefits of technology and mass production. They bought new homes, automobiles, televisions, and other consumer goods – sustaining the expansion of the domestic economy.[1]

During this period the federal government increased its control particularly over interstate industries, and the economic performance of these regulated industries was as impressive as that of nonregulated business. The modern airline industry emerged under the guidance of the Civil Aeronautics Board (CAB) and the Federal Aviation Administration (FAA); the Federal Communications Commission (FCC) oversaw the develop-

---

[1] Louis Galambos and Joseph Pratt, *The Rise of the Corporate Commonwealth: United States Business and Public Policy in the 20th Century* (New York: Basic Books, 1988), pp. 129–37, 155–8. Also see Harold G. Vatter, *The U.S. Economy in the 1950s* (New York: Norton, 1963); and Harry N. Scheiber, Harold G. Vatter, and Harold Underwood Faulkner, *American Economic History* (New York: Harper and Row, 1976).

ment of the national telecommunications industry; the Interstate Commerce Commission (ICC), which already regulated railroads and oil pipelines, saw its powers extended to include water ways and trucking.[2] The Federal Power Commission likewise oversaw the development of the nation's electric power and natural gas industries. As historian Gabriel Kolko postulated, regulation was often imposed to benefit business, and the experience of the American Era seemed to prove that the new regulated system served both business and consumers equally well.[3]

What was not so apparent to most observers was that the regulatory process had its own special dynamic. Like most cartels, regulated industries seemed always to be challenged by competitors who were either outside the regulatory net or able to evade its controls. Thus, there were calls to extend regulation and to perfect the controls that existed. Stockholders under the regulated regime worked very hard to preserve their protected markets, as MichCon had in defending the Detroit industrial market. Regulation was more oriented to stability and equity than it was to innovation and efficiency. It was often framed in response to relatively short term economic developments and had difficulty coping with long-term considerations. These patterns of regulatory evaluation can be seen in the histories of the Federal Power Commission, the natural gas industry, and of Panhandle Eastern. Maguire's company would be involved – directly in some instances, indirectly in others – in many of the major events in those complex histories.

Perfecting the Natural Gas Act

The Natural Gas Act provided a set of guidelines for industry expansion and operation, but it contained certain phrases which generated divergent interpretations and considerable confusion. One of the first significant problems stemming from the act involved section 7(c). In its original form, this provision limited the FPC's jurisdiction "to facilities for the transportation of natural gas to a market in which natural gas is already

---

[2] Galambos and Pratt, *The Rise of the Corporate Commonwealth*, pp. 142–53.
[3] See Gabriel Kolko, *The Triumph of Conservatism: A Reinterpretation of American History, 1900–1916* (New York: Free Press of Glencoe, 1963).

Table 6.1. *Panhandle Eastern Corporation: Natural gas sales,*
1950–1969 (Bcf)

YEAR	For resale						Direct		Total
	TX, OK, KS	MO	IL	IN	MI	OH	Direct Indust.	Other[a]	
1950	1	8	22	35	78	30	21	9	205
1951	4	10	26	43	90	42	26	11	251
1952	2	11	30	48	110	73	18	30	321
1953	2	11	35	54	116	57	20	19	315
1954	1	12	38	57	116	55	31	10	319
1955	1	13	44	62	131	59	34	8	352
1956	1	13	48	62	143	85	38	10	400
1957	2	16	51	66	142	84	39	14	412
1958	2	18	54	69	140	81	35	14	412
1959	2	18	59	74	148	77	39	14	431
1960	21	[b]	63	81	112	85	40	9	448
1961	22	[b]	65	86	115	88	41	15	502
1962	24	[b]	66	99	127	93	46	22	578
1963	31	[b]	72	105	161	110	51	14	662
1964	16	24	113	123	234	111	61	19	701
1965	17	25	120	131	256	111	67	29	755
1966	17	27	132	143	280	120	73	38	829
1967	13	28	150	154	313	122	69	53	903
1968	12	32	163	164	341	147	67	47	973
1969	12	34	173	178	380	153	66	37	1,033

[a]Other includes export sales to Canada, and field sales
[b]1960–3: sales amount for Missouri included with Kansas.
Note: Numbers may not add up due to rounding.
Source: Panhandle Eastern Pipe Line Company, *Annual Report*, 1951, 1953,
1955, 1957, 1959, 1961, 1963, 1965, and 1969.

being served by another natural gas company."[4] If literally
translated, this meant the FPC had no authority to issue certificates to pipelines proposing to serve markets only served with
manufactured gas or markets not served by any gas. After
much protest by gas as well as coal companies, Congress modified section 7(c), and granted the FPC authority over all interstate gas facilities in all markets. The new provision more
clearly defined the FPC's jurisdiction, and it also (and not inci-
dentally) augmented the agency's powers. Moreover, it signifi-

[4] 52 U.S. 825.

cantly increased the Commission's work load as certificate applications inundated its office.[5] This too was a normal consequence of the regulatory dynamics.

Panhandle Eastern president Edward Buddrus remarked on the difficulties this created from the perspective of an expanding pipeline system. As Buddrus observed, "for one reason or another, the certificate hearings are now a major legal proceeding involving the making of a tremendous record, oral arguments, briefs, formal opinions, and orders of the Commission." Exasperated with the regulatory process, Buddrus noted that "the delay in securing a certificate has resulted in higher construction costs because of the time lapse during which costs increase materially."[6]

There were other problems as well – one of them involved the price-setting formula for natural gas. The U.S. Supreme Court (January 3, 1944) attempted to standardize pricing when it ruled in the Hope Natural Gas Company case that the FPC could not use a pipeline firm's reproduction costs as the basis for determining rates. Instead, the FPC had to use the original cost of the pipeline's facilities, less depreciation, as the base for determining "just and reasonable" rates.[7] The gas pipeline industry opposed this ruling since the monetary value of original cost, less depreciation, would generally be lower than that of reproduction costs; as a result, gas prices would be lower and pipelines would make less money. Moreover, the ruling had the effect of discouraging pipelines from producing natural gas since it would apply a cost of finding (plus depreciation) formula to gas reserves instead of a market-based price.

FPC regulations allowed pipeline firms an approximate 6 percent return, based on the depreciated original cost of the pipeline facilities. In effect, this meant that a gas firm had to continue expanding in order to earn enough money to pay for

[5] Federal Power Commission, *Administration of the Certificate Provisions of Section 7 of the Natural Gas Act* (Washington: GPO, no date), pp. 1–8. U.S. Congress, House Report no. 1290 on H.R. 5249, 77th Cong., 1st sess., pp. 2–3. Also see Castaneda, *Regulated Enterprise*, pp. 45–7.

[6] Federal Power Commission, *Administration of the Certificate Provisions of Section 7 of the Natural Gas Act* (Washington: GPO, no date), pp. 13.

[7] 320 U.S. 591. (*FPC v. Hope Natural Gas Company*). See Vietor, *Energy Policy in America since 1945*, p. 72; and Alfred Kahn, *The Economics of Regulation: Principles and Institutions* (New York: John Wiley and Sons, 1970), vol. I, p. 42.

operations. As a pipeline's assets depreciated, its rate base would vanish (ie., "the vanishing rate base") unless it added new value through expansion. Otherwise, a firm could eventually find its assets completely depreciated and, theoretically, incapable of generating income. Maguire adamantly disapproved of the FPC's pricing and rate-of-return regulations, and he was quick to respond to the FPC's enlarged powers. Recognizing that they needed effective political spokesmen, he and Edward Buddrus helped organize a trade association to represent the interests of the natural gas pipeline industry. Buddrus invited twelve executives from major gas pipelines and producers to meet in Kansas City one week after the Supreme Court's decision in the Hope case. Buddrus, as the organizer, became the first president of their new organization, the Independent Natural Gas Association.[8]

The organization became even more important after the FPC launched (1945) its two-year Natural Gas Investigation. Both Congress and the FPC took a critical look at the Natural Gas Act and its effects during the previous ten years. The investigation solicited industry input, which Buddrus and the Association were pleased to supply.[9]

What Buddrus wanted was market-based, not regulatory pricing: "We have a consuming public which wants to buy our commodity, and our answer is that we are ready, willing, and able to sell such commodity just as any merchants on the main street of the average town of the United States." There should, he thought, be no restrictions on the sale of gas.[10] But that was an unrealistic goal in the current context; instead the regulation of gas production emerged as the most pressing issue during the late 1940s and 1950s.[11] Unclear wording in the act left

---

[8] Another account has the first meeting in St. Louis. This organization was later renamed as the Interstate Natural Gas Association (INGAA).

[9] See Federal Power Commission, *Natural Gas Investigation – Nelson Lee Smith and Harrington Wimberly Report* (Washington, D.C.: GPO, 1948), hereafter referred to as Smith and Wimberly Report, FPC; and Federal Power Commission, *Natural Gas Investigation – Leland Olds and Claude Draper Report* (Washington, D.C.: GPO, 1948), hereafter referred to as Olds and Draper Report, FPC.

[10] "E. Buddrus Goes on the Air," *Panhandle Lines*, October 1945, p. 3.

[11] For several lesser knowr attempts to amend the Natural Gas Act during this time see Folder: "Natural and Manufactured Gas," National Security Resources Board, Central Files, July 1949–April 1953, RG 304, NARA.

open the possibility that it applied to gas production as well as transmission. Several congressmen combined forces to draft the Rizley–Moore and Kerr–Harris bills in an effort to explicitly limit the FPC's regulatory powers over gas production. The legislators wanted to settle the fears of producers and gas firms that owned production facilities.

Neither measure was successful. The House passed the Rizley–Moore bill, but it died in a Senate committee.[12] Both Houses passed the Kerr–Harris bill, which had originated from a suggestion by a Phillips Petroleum Company lawyer to Senator Robert S. Kerr (a founder of Kerr–McGee Oil Company). This measure had the singular purpose of exempting independent producers from FPC wellhead price regulation. But, President Truman – heeding the demand that consumers, not producers, benefit from the regulatory process – vetoed the bill on April 15, 1950.[13] During these debates Panhandle Eastern's management vigorously lobbied to exempt company-owned production from federal regulation.[14]

Developing unregulated reserves

Panhandle Eastern's first substantial effort to integrate upstream into production had occurred near the pipeline's origin

[12] Subsequently, Senator Moore accused FPC Commissioner Leland Olds of masterminding the defeat of his co-sponsored bill. Olds, Moore charged, desired to expand the jurisdiction of the FPC to natural gas production. In what became the beginning of an industry-wide assault on Olds, Moore outlined in a congressional speech Olds's "previous official connections with communism" including a stint as editor of the Federated Press and temporary chairman of the American Labor Party. The Red Scare, fueled by Senator Joseph McCarthy's charismatic anticommunist crusade influenced the natural gas regulatory system as the Senate later voted against Olds's renomination for a third term as FPC commissioner despite Truman's continuing support of Olds. *Congressional Record*, Senate, June 17, 1948, p. 8821.

[13] Richard H. K. Vietor, *Energy Policy in America since 1945*, pp. 76–82. Also see Legislative Reference Service, "The Kerr Bill," May 2, 1950, Folder: "Natural and Manufactured Gas, J3," National Security Resources Board, Central Files, July, 1949–April, 1953, RG 304, NARA.

[14] U.S. Congress, House, Interstate Commerce Committee, Subcommittee on Petroleum and Federal Power, *Natural Gas Amendments* (Production and Gathering) 81st Cong., 1st sess., April 1949, p. 453. During the hearings held in May through June, Panhandle Eastern expressed reservations about legislative attempts to exempt only independent producers.

Figure 6.1 Liberal, Kansas compressor station

in the Hugoton Field. This was the supply source of the firm's "West End" at Liberal, Kansas. Liberal was the epicenter of that vast field, and Panhandle had opened an office there on January 1, 1946.

Maguire was determined to press forward with production in this field, despite his realization that companies with such properties were disadvantaged because of the Supreme Court's decision in the Hope case. Now company-owned gas production property, although a depleting asset, would be treated like all other tangible physical assets. Why, Maguire reasoned, should Panhandle Eastern own gas reserves if the gas could not be produced and sold at market prices? The answer was that it should not and that the production had to be spun off into a separate enterprise.

Company engineer Charlie Hinton assisted in the forthcoming spin-off of these reserves, a substantial portion of which he had helped Panhandle acquire. During the mid-1940s, he had negotiated with Cities Service Oil Company in Bartlesville, Oklahoma, for leases in the Hugoton Field of Kansas. During these discussions, Hinton noticed on a Cities Service map margin a reference to a proposed $.10 per Mcf gas sales contract. Since Hinton knew of no customers paying more than $.08 per

Mcf, he asked who would pay so much. A Cities Service landsman told him: Kansas Power & Light Company.[15]

Hinton immediately telephoned Maguire in New York. He explained that they could sell Hugoton gas produced in Kansas directly to Kansas Power & Light. This intrastate sale would be exempt from FPC regulation. Maguire phoned Mr. Dean Ackers, chairman of the board of Kansas Power & Light Company (and an old associate), and launched negotiations. Kansas Power & Light would construct a nine-mile line south of Ulysses, Kansas, to a predetermined delivery point and pay Panhandle $.12 per Mcf.[16]

To handle this contract without placing the Hugoton production assets in Panhandle's rate base, Maguire organized the Hugoton Production Company (September 22, 1948). Kansas Power & Light would thus purchase its gas in an intrastate transaction not subject to FPC regulation.[17] Panhandle simply transferred to Hugoton all of the gas leases located on ninety-six thousand acres in Grant and Stevens Counties, Kansas, and $675,000 in cash. In return, Hugoton Production transferred all of its outstanding capital stock to Panhandle and gave the parent firm the option to purchase all or part of its gas on or after January 1, 1965. Mo–Kan, still under Maguire's control and owning a substantial amount of Panhandle stock, received approximately 13 percent of Hugoton's stock. Also, Panhandle declared a dividend of one-half share of Hugoton stock for each share of Panhandle Eastern common stock.[18]

The new firm, ostensibly outside the FPC's regulatory system, quickly attracted the attention of the Commission. Here at the microlevel was the problem that would return at the macrolevel to plague the industry, the agency, and the nation. But of course the Commission, attuned to its own dynamics, chose not to address that situation. Two weeks after Panhandle transferred the leases to Hugoton, the FPC ordered an investigation of the transaction. The FPC demanded the immediate suspension of the deal. Panhandle, however, continued with its

---

[15] C. H. Hinton, "Hugoton Production Company," unpublished article, PECA.

[16] Ibid.

[17] 6 FPC 78. Also see William G. Maguire, interview by Robert B. Eckles, August 7, 1956 and June 1 and 2, 1958, PECA.

[18] 93 U.S. L ed 1502–3. Also see T. Boone Pickens, Jr., *Boone: T. Boone Pickens, Jr.* (Boston: Houghton Mifflin, 1987), p. 73.

plans, and the FPC responded by filing suit against Panhandle and Hugoton in the U.S. District Court of Delaware. Although the court issued a temporary restraining order to Panhandle, an appeal to the Third Circuit Court of Appeals went in Panhandle's favor.

The resulting court battles addressed the FPC's authority over gas production. Significantly, the Court of Appeals noted that the Natural Gas Act excluded " 'the production or gathering of natural gas' from the Commission's jurisdiction, left the transfer of gas leases to state regulation and outside the scope of the Commission's regulatory powers." The U.S. Supreme Court reviewed the decision in regard to "the applicability of the Natural Gas Act to this transaction," concluding that the Third Circuit Court and Panhandle had correctly interpreted the law.[19]

Justice Reed pointed to the source of the problem (and to the dynamic of the regulatory process). He said that the FPC's opposition to the Panhandle Eastern–Hugoton deal was not based on a strict interpretation of the act. The FPC argued that the leases were "dedicated to the discharge of Panhandle Eastern's public-utility obligation" and that the Commission's approval was necessary before Panhandle could make the transaction. In disagreement, the majority of the Court concluded that the FPC did not have authority over Panhandle's production facilities: "To accept these arguments springing from power to allow interstate service, fix rates, and control abandonment would establish wide control by the Federal Power Commission over the production and gathering of gas. It would invite expansion of power into other phases of the forbidden area."[20]

Justices Black, Douglas, and Rutledge dissented. Justice Black, who wrote the dissenting opinion, argued that the leases had already been calculated as part of Panhandle Eastern's rate base and could not therefore be separated from the firm without FPC approval. The dissenting judges warned that "The Court's judgment and opinion in this case go far toward scut-

---

[19] 93 U.S. L ed 1501–2. Also see Panhandle Eastern, *Minutes*, December 15, 1948.

[20] 93 U.S. L ed 1506–7. Also see FPC v PEPL, U.S. Dist. Court Delaware, November 13, 1948; FPC v. PEPL, U.S. Court of Appeals, 3rd Circuit, no. 98–17, January 9, 1949.

Figure 6.2 Houstonia compressor station

tling the Natural Gas Act."[21] Since neither Congress nor the FPC sought that outcome, the national debate about the Natural Gas Act continued to heat up during the early 1950s. For the moment, however, Maguire and Panhandle had proven their point and moved one significant part of their operations outside of FPC control.

Return to Detroit

These issues of market versus regulatory control were also being played out in Detroit. The city's council members were not enthusiastic about the fact that Phillips Petroleum was now Michigan–Wisconsin's sole supplier of natural gas. Since Phillips' gas sales to the Mich–Wisc pipeline were unregulated, Phillips in effect was in control of the price at which

[21] 93 U.S. L ed 1511.

MichCon purchased gas from its Mich–Wisc affiliate for resale to Detroit's residential, commercial, and industrial customers. The council members were not at all pleased that Phillips would determine the price of gas for end users in the Detroit area.

Just as these concerns were mounting, Phillips unwisely flexed its monopolistic muscle. Construction of the Mich–Wisc line had proceeded more slowly than planned, causing the pipeline firm to renegotiate its gas purchase contract with Phillips. Phillips forced Mich–Wisc to accept a 70 percent increase in its gas purchase price in exchange for a six-month delay in deliveries. The contract extension was vital to the pipeline firm, and Phillips's position as the pipeline's sole supplier gave it all the leverage it needed to impose this new contract on Mich–Wisc (and thus on Detroit).[22]

Unwilling to abide by this settlement, the City of Detroit filed a motion with the FPC to have Phillips declared a natural gas company subject to the agency's rate regulation. Wisconsin officials were also concerned, and the FPC agreed to study the Phillips situation. In the fall of 1948, the agency ordered an investigation of Phillips's rates and operations. This was an important inquiry, because Phillips was one of the largest gas producers in the U.S. In 1950, it sold 60 percent of its natural gas to five pipelines, Mich–Wisc, Panhandle, Independent Natural Gas Company, El Paso Natural Gas Company, and Cities Service Gas Company. These sales accounted for about 15 percent of all U.S natural gas transported in interstate commerce.[23]

In both Wisconsin and Detroit, important public officials were disturbed by the power Phillips had over gas prices.[24] They were also worried about the political influence Phillips and the other companies in this industry wielded. A *Detroit Free Press* article listed the many prominent political figures who were said to be on the payrolls of MichCon or Panhandle Eastern. The list for MichCon included: Prentis Brown, a former U.S. Senator from Michigan; Murray D. Van Wagoner,

[22] 10 FPC 317.

[23] Ibid., pp. 313–17.

[24] Vernon W. Thomson (Wisconsin Attorney General) to Alexander Wiley (Wisconsin senator), March 13, 1951, Folder: "PE–MC" Data, Box 189B10, PECA.

former governor of Michigan; William J. McBreasty, former chairman of the Michigan Public Service Commission; and Donald Richberg, former director of the National Recovery Administration, among others. Panhandle Eastern's list included William E. Dowling, former Wayne County prosecutor and Detroit Corporation Counsel; John S. L. Yost, who had worked for the antitrust division of the Department of Justice; Leslie T. Fournier, an economist who had worked for the SEC on utility policies; Robert Patterson, former Secretary of War; and Robert Morgenthau (one of Maguire's most trusted counselors), son of FDR's Secretary of the Treasury.[25]

Against this background of political influence, charge and countercharge, the Commission went ahead with its study. The results were disappointing to the critics of the corporations. After a long series of hearings, the FPC, in a 4 to 1 vote, decided (August 16, 1951) that Phillips was not a natural gas company as defined by the Natural Gas Act. The Commission terminated the investigation of the "jurisdictional status and rates of the company."[26] In a parting shot at Phillips, the Commission encouraged dissatisfied intervenors to seek redress through the judicial system where they might seek a new interpretation of the Act.[27]

That is exactly what the state of Wisconsin did. Wisconsin attorney general Vernon W. Thomson led the charge. He requested help from Wisconsin Senator Wiley to break the Mich–Wisc–Phillips monopoly:

> "It is possible that, with your help and the help of Senator McCarthy, Wisconsin can strike at the root of our troubles and end them successfully much sooner? . . . It looks to me as though a gross deception has been practiced upon the people of Wisconsin by the Michigan–Wisconsin Pipe Line interests. . . . Essentially, the fact as I see it is that Wisconsin is being deprived of its right to natural gas at a fair price, not only because of the tight pipe line monopoly here, or because of a seeming lack of arm's length

25 Ralph Nelson, "Gas Firm's Buying Influence with Jobs?," *Detroit Free Press*, May 10, 1951.

26 10 FPC 246. Also see *Congressional Record*, Senate, April 27, 1951, pp. 4574–6 and "Four Pending Gas Cases Affect Wisconsin Users," *The Milwaukee Journal*, December 16, 1951; and "State Again Jostled on Gas," *The Milwaukee Journal*, December 9, 1951.

27 10 FPC 246.

dealing for natural gas from the Phillips interests, but principally because they [MichCon interests] are using the gas consumers of Wisconsin as mere tools or weapons in their fight to stand off competition ". . . especially at Detroit."[28]

Thomson never let up. He demanded that Phillips be subjected to rate regulations. He continued to look for relief to the courts and to Congress. As he explained to the mayor of Cleveland: "An investigation by this office of Wisconsin's natural gas shortage is revealing a need for an amendment to the Natural Gas Act for the purpose of eliminating monopolistic practices in the production and transmission of gas." He would seek a change to federal regulations for the benefit of his state.[29]

Wisconsin won the first two rounds in the struggle. The Court of Appeals for the District of Columbia ruled (May 1953) that Phillips was a natural gas company, and the U.S. Supreme Court at first declined to review the ruling. On January 18, 1954, however, the high court reversed itself and agreed to review the decision. Industry as well as the FPC applauded this new decision. "The appeals court's ruling was a horrible decision," an FPC official declared, "because it didn't give us any guide posts. Now we can hope for a ruling that'll give us some ground rules in the event we still end up regulating these sales."[30]

Wisconsin and the FPC scored a knockout in June 1954, when the Supreme Court reaffirmed the decision.[31] This landmark case – the famous "Phillips Decision" – empowered the FPC to set the price of natural gas sold by producers to interstate pipelines. In one respect, the logic was sound. Since Congress had authorized the FPC to set "just and reasonable rates" with the end users in mind, the logic of control seemed to require that it

---

[28] Vernon W. Thomson to Alexander Wiley, March 13, 1951. Folder: "PE-MC" Data, Box: 189B10, PECA.

[29] Vernon W. Thomson to Thomas Burke, May 1, 1952. Folder: Texas–Michigan Pipeline Company, Box 189B6, PECA.

[30] "Supreme Court to Decide Whether 'Independents' Come under FPC," *Wall Street Journal*, January 19, 1954. Angus Deming, "FPC Decision in Key Gas Rate Case Soon," *Wall Street Journal*, March 30, 1954.

[31] *Phillips Petroleum Company* v. *State of Wisconsin*, 374 U.S. 672. Also see Raymond N. Shibley and George B. Mickum, III, "The Impact of Phillips upon the Interstate Pipelines," *Georgetown Law Journal*, vol. 44, no. 4, (June 1956), p. 628.

do so by regulating the wellhead price of natural gas. The dynamic of regulatory evaluation worked this way, driving the process of control further into those elements of the industry's setting where market forces were still at work. Neither the Court, nor the Commission, nor Wisconsin's officials seemed worried that this extension of price control might in the long run bring about unanticipated results detrimental to the end-users whose interests they were seeking to protect in the short run.

Panhandle Eastern saw this rather clearly. The firm responded to the Phillips decision with this statement: "We believe this development may seriously curtail the supply of natural gas that will be made available to interstate pipe lines."[32] Panhandle, as well as other gas firms, issued urgent calls for Congress to pass legislation exempting producers from regulation. As they pointed out, the Federal Power Commission had jurisdiction over 157 natural gas companies prior to the Phillips decision. After the decision, the agency had to regulate more than 4,365 independent producers who sold natural gas to interstate pipeline companies. This, the companies thought, would be an impossible task.

At the same time that the U.S. Supreme Court was deliberating the Phillips case, the FPC was considering a Panhandle Eastern case which might have mitigated the adverse effects the Phillips decision was likely to have on the supply of gas for interstate shipment.[33] Panhandle, which now sold gas to forty-nine utility customers, had sued the FPC to allow it to use the "fair field" price for company-produced gas, the price based upon other gas sold from the same field, instead of using the depreciated original cost rate base method. Although the price increase to individual customers would be small, the fair field price formula would give Panhandle a $21.4 million rate boost. The FPC agreed with Panhandle, and adopted as its formula (April 15, 1954) "a price reflecting the weighted average arm's length payments for identical natural gas in the fields (and sometimes from the very same wells) where it is produced. . . ."[34] In this same case, the FPC allowed Panhandle Eastern a 5¾ percent

[32] Panhandle Eastern, *Annual Report, 1954*, p. 5.

[33] Richard H. K. Vietor, *Energy Policy in America since 1945* (Cambridge University Press, 1984), p. 83. Also see Panhandle Eastern, *Annual Report, 1954*, p. 11.

[34] Vietor, *Energy Policy in America since 1945*, p. 83. Also see 13 FPC 53–4 and 13 FPC 59–77.

rate of return (as opposed to the 5 ½ percent requested by the City of Detroit and FPC staff and the 6 ½ requested by Panhandle).[35]

If sustained, this policy would probably have ensured that sufficient gas would continue to flow into interstate pipelines. Panhandle Eastern, under increasing rate reduction pressure, had notified its stockholders as early as 1952 that there existed "a growing scarcity of natural gas relative to the increased demands. Your company grows more reluctant to spend the large sums of money required to acquire new reserves in view of the Commission's persistent limitation of the Company's return on investment in production to a small fixed per cent."[36] The "fair field" ruling changed that situation, and Panhandle's management responded positively: "now [Panhandle] has the necessary encouragement and incentive to increase substantially production of its own gas reserves and exploration for new reserves."[37]

Those political leaders most concerned about the immediate interests of consumers were of course chagrined by the decision. In Congress, Senator Lee Metcalf (D–MT) said "the fair field price is the highest monopoly price the traffic will bear." He went on to lambast the entire regulatory process: "Far better we should discontinue the pretense that the FPC is regulating natural gas rates. The consumer then would not be suffering the delusion that he is being protected against exploitation at the hands of natural gas companies."[38] The City of Detroit filed suit against the FPC opposing the higher cost.

[35] 13 FPC 95 and 100.

[36] Panhandle Eastern, *Annual Report, 1952*, p. 14.

[37] Panhandle Eastern, *Annual Report, 1954*, p. 3. Although the FPC allowed Panhandle Eastern only a 5 ¾ percent return, Panhandle management appeared willing to go along with the FPC's decision even though it meant that for the period during which Panhandle's rates based on a 6 ½ percent return were in effect from February 20, 1952, through May 1, 1954, Panhandle Eastern would have to refund as much as $32 million to customers. Since Panhandle had been required to establish an escrow account when it began charging the new rate, the account held the money needed to make the refund. In a related ruling regarding the primacy of state minimum price regulation, the FPC ruled that it did not have to abide by state minimum prices. Panhandle Eastern, enjoying some of the minimum prices in states such as Oklahoma, objected to these rulings. Panhandle Eastern, *Annual Report, 1952*, pp. 15 and 18.

[38] *Congressional Record*, Senate, 83rd Cong., 2nd sess., vol. 100, no. 57 (March 29, 1954).

On December 15, 1955, the U.S. Court of Appeals, D.C. Circuit, overturned the decision. The court ruled that "the allowance of a field price should not lift the rates above the 'just and reasonable' standards of the Natural Gas Act."[39] As Richard Vietor concluded: "Had this decision to allow a fair field price withstood judicial review, it would have meant that FPC regulation of independent producers under the Phillips doctrine could have proceeded with little disruption."[40] But that was not to be the case; Panhandle's appeal was unsuccessful.[41] The fair field doctrine was thrown out, and with it, the opportunity for firms in the industry to sustain development over the long term.

In one last ditch effort to release independent producers from regulation, Senator J. William Fulbright (D–AR) and Representative Oren Harris (D–AR) introduced the Fulbright–Harris bill. The bill would replace the "cost-based" method of determining prices with the "fair field price" which the FPC had earlier approved. This bill would therefore overturn the Court decision, in the 1954 Panhandle case. But on the very day the Senate voted, Senator Francis Case (R–ND) informed his colleagues that an oil industry lobbyist had attempted to bribe him. From the point of view of President Eisenhower, these charges tainted the bill. The Republican administration favored the measure, but after both the House and Senate passed it, Eisenhower vetoed the bill. He would not, he said, sign a bill that possibly was tainted with bribe money.[42] Eisenhower noted that the legislation was the type "needed because the type of regulation of producers of natural gas which is required under present law will discourage individual initiative and incentive to explore for and develop new sources of supply."[43] But subsequent efforts to pass the necessary legislation were unsuccessful.

---

[39] 230 F. 2(d) 810.

[40] Vietor, *Energy Policy in America since 1945*, p. 84.

[41] Panhandle Eastern, *Annual Report, 1956*, p. 5.

[42] Vietor, *Energy Policy in America since 1945*, pp. 87–9. A Senate investigation of the alleged bribe revealed that Superior Oil Company had paid Senator Case $2,500. The oil company later paid a $10,000 fine for not registering its lobbyists. See David Howard David, *Energy Politics*, 2nd ed. (New York: St. Martin's Press, 1978), pp. 122–3.

[43] "Management's Report on the Gas Bill Veto," *Panhandle Lines*, February 1956. Also see "Presidential Message, February 17, 1956," Box 726, file 140-C, White House Official File, Dwight D. Eisenhower Presidential Library.

Figure 6.3 Trunkline Gas Company control room

Maguire reaffirms corporate control

During this era of regulatory turmoil and intense competition for gas sales, Maguire tightened his grip on the internal management of Panhandle Eastern. He had relied on company president Ed Buddrus to run the day-to-day operations from Kansas City, while Maguire reigned on Wall Street. Buddrus complained to Maguire during one of their regular New York luncheons in 1953 that "nobody pays attention to my ideas. . . . I might as well quit." Maguire snapped back that the next time he said that, the offer would be accepted. Eight months later at the same restaurant table, Buddrus complained again. Maguire accepted his resignation. The board then elected Maguire, already Panhandle's CEO and chairman, as president. The board also elected Hy Byrd, president of Trunkline, to be the executive vice-president of Panhandle Eastern.[44]

Maguire soon lost Byrd, as well, after a dispute over Pan-

[44] Raymond M. Shibley, interview by Christopher J. Castaneda and Clarance M. Smith, January 13, 1994, PECA.

handle Eastern's involvement with a new independent effort
to construct a pipeline from the Gulf Coast of Louisiana to
West Virginia, where it would connect with United Fuel Gas
Company, a subsidiary of Columbia Gas. Hy Byrd actively
pursued the acquisition of Gulf Interstate. He purchased on
behalf of Panhandle a one-half interest in the venture and
planned its integration into the Trunkline system.[45] After subsequent discussions, including disagreements over financing
arrangements between the two firms Trunkline canceled the
agreement. Byrd, angered by Maguire's opposition to the transaction, left Trunkline (along with several other key employees) to become president of Gulf Interstate (later renamed as
Columbia Gulf Transmission). Maguire then assumed Byrd's
position as president of Trunkline.[46]

In a twelve-month period, Maguire had driven off two of
his top executives. Nevertheless, he seemed to thrive in such
situations, often pitting his vice-presidents against each other
to spur their performance (Table 6.2). He continued to use
the corporate headquarters in New York as his base, and he
dominated his highly centralized operations by rarely dele-
gating significant decision making power to subordinates.[47]
As he looked again to the Detroit market and his rivals at
MichCon, very little had changed. His old opponent William
Woolfolk had died in 1953 and been replaced by forty-seven-
year-old Ralph T. McElvenny, a Stanford law school graduate.[48] But Maguire's strategy and tactics were unchanged as
he tried again to achieve a stronger position in this lucrative
market.

---

[45] The purchase was made on March 10, 1952.

[46] Trunkline, *Minutes*, April 20, July 7, September 28, and October 19, 1953.
Gulf Interstate received a certificate from the FPC on May 20, 1953. After
the line was completed in 1954, Gulf Interstate served as a contract carrier
for the Columbia System, much as Trunkline served as a virtually exclusive
carrier for Panhandle Eastern. In 1958, Columbia purchased the Gulf Interstate line and renamed it Columbia Gulf Transmission Company. 12 FPC
116; and Tussing and Barlow, *The Natural Gas Industry*, p. 218.

[47] "W. G. Maguire Named Company President," *Panhandle Lines*, March 1953,
p. 3. Also see Robert VanLeuvan, interview with Clarance M. Smith and
Christopher J. Castaneda, March 16, 1993, PECA.

[48] "Detroiter Spends $250 Million to Bring Natural Gas Home," *Detroit News*,
October 23, 1955.

Table 6.2. *Panhandle Eastern Pipe Line Company: Earnings and financial data, 1948–1968*

| | | Net income | |
| | Operating revenue (millions) | Operating income (millions) | Retained by business (millions) |
Year			
1948	$34	$ 9	$ 2
1950	41	11	2
1952	77	15	3
1954	82	14	4
1956	113	20	9
1958	121	19	7
1960	151	25	8
1962	210	38	12
1964	247	42	14
1966	285	49	15
1968	337	59	17

Source: Panhandle Eastern Pipe Line Company, *Annual Report*, 1969, 1960, 1958, and 1957.

Another round against MichCon

At this time, MichCon needed even more gas. Rather than renewing the Panhandle Eastern contract, however, the utility's holding company proposed to build a second supply line, the American Louisiana Pipeline. This 1,200 mile addition would extend from the Louisiana coast to the Midwest and deliver natural gas to Detroit and other midwestern cities; it would provide access to gas supplies in the Gulf of Mexico.[49]

Panhandle viewed this proposal as another threat (even though Panhandle was then attempting to stop delivering gas to MichCon). In an attempt to persuade MichCon officials not to proceed, Maguire invited MichCon's president and general counsel to New York to discuss the Detroit gas situation. The meeting lasted all day. Maguire said that he was now prepared to work with MichCon's holding company, American Natural, to settle the existing controversies. He also made it clear that he was very concerned about the proposed American Louisiana

[49] "Dwellers Dig for Treasure," *Detroit News*, January 24, 1956.

pipeline. That line, he said, would be very costly and would be a high-cost supplier. Maguire proposed instead that Panhandle and Trunkline increase their capacity through a combination of additional looping and compressor stations: they would sell the gas to MichCon on favorable terms. In the meantime, the two companies would have to settle their outstanding differences including a lawsuit in which Panhandle charged that MichCon owed it $2 million.

Attorney Daniels drafted a memorandum of record after the meeting:

> I distinctly had the impression when they left that they were sincerely interested in the possibility of an over-all adjustment and in abandoning their ideas for the American-Louisiana pipe line if such an adjustment could be worked out. At no time during the entire conversation did they say, or even suggest, that the American Natural Gas Company interests were not willing to consider such an adjustment; nor did they say that plans for the American-Louisiana pipe line had gone so far that American Natural's steps in that connection could not be retraced. In a word, I had the distinct feeling when they left that real progress had been made during the day in the direction of an over-all adjustment of existing controversies.[50]

As it turned out, however, none of the problems were resolved. MichCon's president, Henry Tuttle, wrote Maguire: "As a result of our meeting . . . we have concluded that our present and future gas requirements will best be served by an increase in the gas supply of the American Natural Gas System through the construction of a pipeline by the American Louisiana Pipe Line Company."[51] Maguire's response was succinct: "The basic dispute between our companies stems from our belief in competition and your wish for a monopoly in all types of gas service despite the fact that you have no franchise in the Detroit area."[52] Maguire accused the MichCon organization of attempting to establish an Insull-like monopoly in the Midwest. After the FPC approved the American Louisiana certificate applica-

[50] Memorandum by Joe J. Daniels, October 26, 1953, Folder: "1953: PE–MC Letters," Box 189B21, PECA.

[51] Henry Tuttle to William G. Maguire, September 17, 1953. Worth Rowley Box, PECA.

[52] William G. Maguire to Henry Tuttle, October 12, 1953. Worth Rowley Box, PECA.

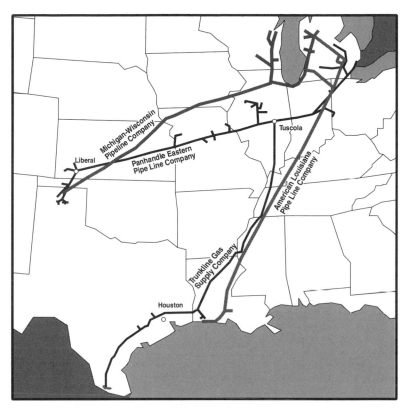

Map 6.1 Michigan–Wisconsin Pipeline Company and American Louisiana Pipe Line Company, 1958

tion, Panhandle filed suit in the U.S. Circuit Court of Appeals in Philadelphia to prevent the construction of the new line. But the court said that since there was a great need for additional gas in the area and Panhandle had not filed a competitive application, it was denying Panhandle's request.[53]

Even though the American–Louisiana line was completed, MichCon continued to need Panhandle's gas, and the FPC continued to compel Panhandle to deliver gas to the utility (Map 6.1). For about eight years after the expiration of Panhandle's contract with MichCon, the FPC refused to allow Panhandle to

[53] "Court Backs New Pipeline for Michigan," *Detroit City News*, February 26, 1955.

stop delivering gas. Michigan Consolidated, with 875,000 customers, remained one of Panhandle Eastern's biggest volume purchasers. Even though the two firms were locked in a seemingly everlasting struggle, Panhandle was supplying approximately 45 percent of the gas distributed in Detroit.[54]

McLouth Steel

Despite several rebuffs, Maguire was still convinced that he would ultimately be able to bypass the utility and make direct sales to large Detroit customers. For Maguire, "no" always meant "maybe." He now approached McLouth Steel Corporation, explaining that Panhandle would have some excess natural gas available for sale "down river." Panhandle Eastern's interest in this deal increased when company officials learned that General Motors had a $25 million investment in the McLouth plant.[55] In a subsequent meeting, McLouth "brought up the matter of [Panhandle] coming into MichCon's territory and if there would be any adverse publicity in connection with their purchasing gas from us." Maguire's spokesman said Panhandle would need to go before the MPSC for approval and that MichCon could intervene in those hearings.[56]

Panhandle proposed that McLouth construct a line and install a metering station between its plant and Panhandle Eastern's line, a distance of about eleven miles. If Panhandle built the line it would require an FPC certificate, but if McLouth built it, Panhandle would only need a certificate from the MPSC. The steel firm objected to this idea, but Panhandle countered with an offer to provide engineering services and "line up a right-of-way man to get the right of way. . . ." Panhandle also said it would give McLouth a $.05 per Mcf credit on its gas purchase price ($.35 per Mcf) if McLouth would then deed the line and station to Panhandle. On December 28, McLouth sent

---

[54] See "McNamara Rips Ike on Drive to End Gas Price Regulation," *The Detroit News*, March 24, 1955.

[55] George L. Ballard to Fred H. Robinson, August 6, 1953, Folder: "McLouth Steel," Box 189B20, PECA. Also see "Chronology of McLouth Steel Corporation Proceedings"; and F. H. Robinson, *Statement*, August 6, 1956, Folder: "McLouth Steel," Box 189B20, PECA.

[56] George Ballard to Fred Robinson, September 24, 1953, Folder: "McLouth Steel," Box 189B20, PECA.

the "Letter of Intent" necessary to get the process of acquiring Michigan's approval of the line. The two firms signed a contract for industrial interruptible gas deliveries on March 1, 1954.[57]

The following month Panhandle filed an application with the MPSC for a certificate to sell gas to McLouth. Panhandle noted that it currently provided more than 45 percent of the gas used for domestic and industrial purposes in the Detroit area. At the same time, Panhandle issued a news release which attacked MichCon:

> Today's application brings to a head a dispute of almost ten year's duration between Panhandle and Michigan Consolidated over the former's right to sell gas to industrial users. Panhandle officials contend that Congress under the Natural Gas Act gives them the right to compete on an open market and that any infringement of that right constitutes a monopolistic restraint of trade. They further charge that Michigan Consolidated does not have gas in sufficient quantities to supply industrial demands and even if it did have an adequate supply, it could not sell at a price that would compete successfully with other fuels.[58]

Panhandle maintained that its agreement with McLouth would in no way take gas away from Michigan's residential gas customers. Panhandle also pointed out that MichCon's gas sales price to McLouth would be about $.59 per Mcf compared to $.30 per Mcf Panhandle was charging. The advantage to the consumer of direct sales was being able to bypass the distributor's rate base.

MichCon fought this contract as it had all of the others, but in early August, Panhandle learned that the MPSC would approve the contract. The commission was not entirely satisfied with Panhandle Eastern's case and noted that if MichCon sold gas to McLouth, the distributor could then "reduce the price of gas to residential and commercial customers of the

---

[57] George L. Ballard, Memorandum, October 22, 1953; and George L. Ballard, Memorandum, December 4, 1953; C. T. Murphy to Panhandle Eastern Pipe Line Company, December 28, 1953, Folder: "McLouth Steel," Box 189B20, PECA. McLouth initially requested 1 MMcf/d at its Trenton Plant and 2 MMcf/d at its Gibraltar plant with volumes increasing to 6 and 10 MMcf/d, respectively.

[58] Panhandle Eastern, *News Release*, April 5, 1954, Folder: "F. H. Robinson: Detroit Settlement, 1961," Box 189B20, PECA.

company."[59] Panhandle appeared nevertheless to have won the battle. On August 3, 1956, the MPSC voted two to one in favor of Panhandle.

But of course, MichCon challenged the MPSC order in court, and in the meantime, one of the two members of the MPSC who had favored Panhandle, resigned. He was replaced by a commissioner friendly to the utility, and on August 23, the MPSC opened the way for MichCon to sell large volumes of industrial gas to McLouth. MichCon began immediately to deliver gas to the steel firm. Although the case continued for several more years, Maguire had once again lost this battle. It was no easy task for "foreign" firms to gain entry to a market controlled by a local business buttressed by a state regulatory commission dedicated to a short-term concept of consumer interest.[60]

MichCon and Panhandle Eastern continue their struggle

After American Louisiana Pipe Line Company began operations in August 1956, Panhandle again sought FPC permission to abandon service to MichCon. During the course of the next year, the FPC held intermittent hearings which resulted in 12,000 pages of testimony and 259 filed exhibits. Panhandle did not want to sell natural gas at rates 40 percent below those MichCon was paying to its affiliate suppliers, Mich–Wisc and American Louisiana, particularly when it could not bypass MichCon and sell gas directly to industrial customers. On January 7, 1958, the FPC staff recommended to the FPC examiner and the full commission that the abandonment be approved. After the presiding examiner revoked the abandonment, Panhandle appealed to the full Commission. On December 19, 1958, the FPC ruled that MichCon did "not have a vested right to the lower priced Panhandle gas." The agency gave Panhandle thirty days to submit a plan for the reallocation of the current 127 MMcf/d it was delivering to MichCon.[61]

---

[59] Henry A. Montgomery to Trustee, Brownstone Township, August 27, 1956, Folder: "McLouth Steel," Box 189B20, PECA.

[60] See "A Factual History of How One Michigan Industry Was Denied Competitively Priced Fuel," a report prepared by the Detroit Office of Panhandle Eastern, Folder: "F. H. Robinson: Detroit Settlement, 1961," Box 189B23, PECA. Also see 98 N.W. 2d 592.

[61] 283 F. 2d 204 and 267 F. 2d 459. Also see "FPC Approves Abandonment Application," *Panhandle Lines*, January 1959, pp. 12–13. Panhandle East-

It now appeared that Maguire had at last freed himself from MichCon, but of course in regulatory politics it was difficult to reach a firm conclusion in a decade – often such struggles become a lifetime career. In the midst of the conflict over abandonment of service, Panhandle Eastern filed a $601,354.39 claim against MichCon for withholding payment for gas already delivered. MichCon countered with a $2 million suit against Panhandle, but this time the court dismissed MichCon's claim and granted the judgement to Panhandle.[62]

Panhandle moved forward with its effort to discontinue service. MichCon delayed by way of legal maneuvering. But, on October 2, 1959, at 9:00 P.M., Panhandle Eastern finally stopped all gas deliveries to MichCon.[63] Once again, MichCon turned to the Circuit Court of Appeals, which set aside the FPC's order and remanded it to the Commission. On December 19, 1959, the FPC reaffirmed its earlier approval of abandonment, but MichCon filed another motion requesting resumption. Then, in April of 1960, the U.S. Court of Appeals for the District of Columbia again set aside the FPC abandonment decision, and the FPC directed Panhandle to begin delivering 100 MMcf/d to MichCon on an interim basis between September 23 and November 1, 1960.[64]

This time, Maguire agreed to a compromise. In the resulting agreement, MichCon received some but not all of the gas it wanted, and Panhandle could proceed with a major expansion of seasonal service, storing large volumes in Michigan.[65] Nei-

---

ern, *Annual Report 1958*, p. 23. "FPC Approves Abandonment Application," *Panhandle Lines*, January 1959, p. 13. Also see "Panhandle's Plea Rejected," *Detroit Night Times*, April 20, 1955; and "Detroit Wins Natural Gas Supply Fight," *Detroit News*, April 20, 1955.

[62] 173 F. Supp. 738. "Mich–Con Loses, Pays Damages," *Panhandle Lines*, June 1959, p. 15.

[63] "Panhandle Ends Delivery to Mich–Con," *Panhandle Lines*, October 1959, p. 15.

[64] "Mich–Con Deliveries Resumed," *Panhandle Lines*, October 1960, p. 12.

[65] "Joint Panhandle–Trunkline Proposal Would Settle Long Abandonment Case," *Panhandle Lines*, January 1961, p. 4. Also see, "Panhandle offers a Compromise in Detroit Gas Case," *Wall Street Journal*, January 10, 1961, p. 14; and "Panhandle, Michigan Consolidated Wrangle over Gas Supply Ended," *Wall Street Journal*, January 10, 1962. On January 6, 1961, Panhandle and Trunkline filed a joint proposal to provide additional gas to utility firms in six midwestern states. The proposal called for Trunkline to deliver

ther party was entirely satisfied, but at least Maguire and Panhandle could turn their attentions to other matters. Maguire, unlike Woolfolk, lived to see this dispute settled.

Panhandle Eastern rates and federal regulation

Most pressing was the ongoing fight with the FPC over the method of calculating the price of the company's affiliated production.[66] Panhandle asked to reopen its 1955 rate case, and the Commission agreed. Significantly, the agency explained that it was reopening the case to learn if the commodity value formula would save customers money and to encourage the development of additional gas supplies. "Reopening will not only give Panhandle an opportunity to introduce this additional evidence," the Commission said, "it will also provide us an opportunity to consider the augmented record in this case at the same time we are considering the landmark Phillips Petroleum Company independent producer rate case, Docket no. G-1148, et al, which is also before us now for decision."[67]

On September 28, 1960, the FPC issued its decision on the latter case by establishing geographic area prices instead of using the cost-of-service method. The FPC admitted that under area pricing, efficient producers who kept their costs down were penalized with lower prices.[68] Nevertheless, the U.S. Court of Appeals, District of Columbia, affirmed this pricing scheme and was supported by the Supreme Court. As a result,

---

21.4 Bcf annually to American Natural and begin a $77 million expansion program. Panhandle also proposed to deliver large volumes of natural gas to Ford Motor's River Rouge Plant.

[66] Panhandle Eastern, *Annual Report, 1962,* p. 16; and Panhandle Eastern, *Annual Report, 1963,* p. 13. On February 10, 1959, the FPC disallowed Panhandle Eastern's 1955 rate increase implementation and proposed lower rates, once again requiring Panhandle Eastern to pay refunds. In April, the FPC decided against Panhandle's rates for the period covering February 1952 through August 31, 1958. On June 30, 1962, the U.S. Court of Appeals agreed with the FPC. Panhandle then filed a writ of certiorari on the commodity issue with the U.S. Supreme Court, but the Supreme Court denied the filing on February 18, 1963. Panhandle Eastern was thus faced with refunding $53.6 million including interest.

[67] Panhandle Eastern, *Annual Report, 1959,* p. 19.

[68] 24 FPC 547.

Panhandle was forced to refund $54 million collected under fair field pricing.

The new policy proved difficult to implement. The FPC did not issue its first area price until 1965, five years after it announced its adoption of the new method. The Permian basin of Texas received the first area price, but by that time it was obvious that the agency was being overwhelmed by the complexities of the gas industry. Still, the FPC plodded ahead, and the dynamics of regulatory evolution took hold. Each effort to provide legal clarification of technical terminology only served to extend federal regulation into new areas. As the regulatory system became ever more complex, as the process of economic change slowed, as profits were squeezed out of interstate commerce, unanticipated consequences followed.

Gas pipeline firms had strong incentives to invest their capital outside of the regulated natural gas industry; they began to diversify into related, and then unrelated, businesses. Pipelines and producers now had strong disincentives to spend capital on producing natural gas for sale or resale in interstate commerce. All of this was predicted by gas industry operators and observers of the mid-1950s, but the regulators, courts, and legislators pushed ahead with their effort to keep prices low regardless of the long-term effects this might have on the industry and the very consumers they were trying to protect.

7

Strategic diversifications

The New Deal regulatory state broke apart the Power Trust in an attempt to bring competition to the utility industry. As with all regulatory efforts, however, unanticipated consequences emerged. One of these was a widespread movement toward diversification. In effect, regulation drove capital away from the development of this industry, pushing it toward other markets, some of which were closely linked to gas production and distribution, some of which were not. This phenomenon illustrated the tension that was being created by a regulatory system that tended to frame policy on a case-by-case basis in terms of current stakeholders and a business outlook which sought extended profitable investment. Short-term regulatory strategy and long-term corporate strategy dictated different outcomes for the firms involved, the consumers the FPC was trying to protect, and indeed the national economy. The differences would not become obvious, however, until the industry faced a crisis.

Regulation was not the only factor promoting diversification during these years. Throughout the U.S. business system, companies were experimenting with new types of investments in related and often unrelated enterprises.[1] As a tool of business

[1] Between the years 1948 to 1979, U.S. business diversification took place at a high rate, while the highest rates of diversification occurred between the years of 1959 to 1969. Neil A. Fligstein, *The Transformation of Corporate*

development, diversification served many purposes. It allowed firms which served mature markets to repeat their early profitable growth phases, and it allowed firms to exploit more fully their technological and financial resources. Leading companies like Du Pont and General Motors successfully implemented diversification strategies.[2]

To manage diversified operations effectively, most of these firms adopted a multi-divisional structure. In a multi-divisional, or M-Form organization, separate business units were managed individually. "The resulting multi-unit enterprises," economic historian Louis Galambos noted, "developed managerial hierarchies that enabled them to coordinate more efficiently than the market did the complex array of activities they controlled."[3] At Panhandle Eastern, however, William Maguire's impulse to control all aspects of the firm's operations made it difficult for him to decentralize control. Maguire wanted to oversee all of the company's diversification efforts personally. Whatever the structure of the organization, Maguire was going to continue to have a centralized business.

As many of his executive staff members recognized, Maguire was a supreme financial man, but a poor manager. He had a good

Control (Cambridge, Mass.: Harvard Univ. Press, 1990), pp. 263–9. Business historian Alfred D. Chandler, Jr., noted that ". . . diversified, multidivisional enterprises were able to intensify competition within the industry or region into which they moved and at the same time to transfer resources from older, more stable industries or markets to newer, more dynamic ones." Alfred D. Chandler, Jr., *Scale and Scope: The Dynamics of Industrial Capitalism* (Cambridge: The Belknap Press of Harvard University Press, 1990), p. 45.

[2] For an intensive examination of the development of several major American firms see, Alfred D. Chandler, Jr., *Strategy and Structure: Chapters in the History of American Industrial Enterprise* (New York: Anchor Books, 1966). For a more general discussion, see, Louis Galambos and Joseph A. Pratt, *The Rise of the Corporate Commonwealth: U.S. Business and Public Policy in the Twentieth-Century* (New York: Basic Books, 1988), p. 159; Richard P. Rumelt, *Strategy, Structure and Economic Performance* (Boston: Division of Research, Harvard Graduate School of Business Administration, 1974); and Jon Didrick, "The Development of Diversified and Conglomerate Firms in the United States, 1920–1970," *Business History Review*, vol. 46 (Summer 1972), pp. 202–19.

[3] Louis Galambos, "Technology, Political Economy, and Professionalization: Central Themes of the Organizational Synthesis," *Business History Review*, vol. 57 (Winter 1983), p. 473.

eye for financial opportunity, and he was quick to develop those opportunities if they were in areas related to natural gas such as oil production, petrochemicals, and other related chemical ventures. But he tried to manage those undertakings with financial, as opposed to operating goals, in mind. His approach was consistent with the style of management described by Neil Fligstein: "Because financial performance was all that mattered, managers with the financial conception of control pursued growth in whatever industry there were opportunities."[4] Or as an associate of Maguire's explained: "Mac believed that he ought always to make a buck . . . [and] his primary concern was to make a buck as fast as possible."[5] In the postwar natural gas pipeline industry, however, the FPC only allowed firms to make a 6 ½ percent return on their operations – and in some cases, even less. This restriction, among others, trained Maguire's eyes toward new products and activities outside of gas sales.

Maguire and his fellow executives at Panhandle Eastern loudly voiced their criticism of the FPC. Vice-president Fred Robinson, speaking in Kansas City, attacked "the delays caused in the past by such regulation and the restrictive nature of FPC decisions. . . ." He said the agency had failed "to establish and live by a set of rules and regulations upon which the company can make business decisions."[6] As Robinson pointed out, pipeline firms that owned production properties could not be certain of the economic viability of their search for new supplies. The Phillips case and the reversal of the fair field price, along with the recent decisions of the Supreme Court, had left Robinson and others willing to "fight for realistic regulation."[7] Meanwhile he said: "One of the things that has become increasingly apparent to your management is the desirability of separating from the regulation phase of our business all enterprises which can be operated separately."[8] Other voices – some of them from outside the industry – were beginning to express similar concerns. In 1960, James M. Landis, a former SEC commissioner and dean

---

[4] Fligstein *The Transformation of Corporate Control*, p. 226.

[5] Glen Clark, interview by Robert Eckles, January 22, 1957.

[6] "The Future of Panhandle," *Panhandle Lines*, November–December 1961, pp. 12–13.

[7] Ibid.

[8] Ibid.

of the Harvard Law School, explained what was happening to president-elect Kennedy. In Landis' *Report on Regulatory Agencies to the President-Elect*, he described the FPC as the agency which "without question represents the outstanding example in the federal government of the breakdown of the administrative process." The huge backlog of cases filed since the Phillips decision had, he said, literally inundated and paralyzed the agency.[9]

Those firms that wanted to avoid similar paralysis turned to diversification. Tennessee Gas (later named Tenneco, Inc.) ventured into a wide array of activities including farm equipment manufacture and shipbuilding; Mississippi River Fuel Company went into the drilling mud business; Transcontinental Gas (later Transco Energy, Inc.) took a less radical course by retaining a commitment to fossil fuels. Panhandle Eastern's diversification ventures were even more conservative, as they were related directly to its ongoing oil and gas business. Rather than entering entirely new markets, Panhandle sought incremental system expansion which would utilize its existing technologies and natural gas as a base fuel source.[10]

National Petro–Chemicals

The petro–chemical and liquefied petroleum gas (LPG) businesses were closely related technological activities. Both undertakings required the extraction of naturally occurring hydrocarbon liquids from the natural gas stream prior to pipeline transmission.[11] In Houston alone, five liquefied petroleum plants were erected in the wake of the Supreme Court's Phillips decision. The federal regulation of gas producers encour-

[9] James M. Landis, *Report on Regulatory Agencies to the President-Elect*, Senate Committee on the Judiciary, 86th Cong., 2d sess. (Washington: Government Printing Office, 1960), p. 54. For a description of the evolution of Landis' thought see Thomas K. McCraw, *Prophets of Regulation*, pp. 153–209 and 234–5.

[10] Glen Clark, interview by Robert B. Eckles, December 27, 1956, pp. 7–8, PECA. For a discussion of efforts by pipelines to diversify see, "The Gas Pipelines: A New Set of Rules?," *Forbes*, September 1, 1963.

[11] See, Richard H. K. Vietor, "The Synthetic Liquid Fuels Program: Energy Politics in the Truman Era," *Business History Review*, vol. 54 (Spring, 1980), pp. 1–34.

aged the development and promotion of new technologies, fuels, and markets.

As early as 1951, National Distillers & Chemical Corporation and Panhandle Eastern had formed a cooperative venture to construct a petro-chemical facility. The plant was near Tuscola, Illinois, a small town located at the confluence of the Trunkline and Panhandle Eastern systems. Both lines had compressor stations located there, and the Panhandle district office was housed in Tuscola as well. National Distillers retained a 60 percent interest in the venture, leaving a 40 percent share, or a $7.5 million investment, to Panhandle. The joint venture resulted from a series of meetings between Maguire and National Distillers's president, John Bierwirth. Bierwirth was on the Panhandle Eastern board of directors, and Maguire held a similar position at Bierwirth's company.[12]

These two friends and business associates had talked during lunch one day about the possibility of building a petro-chemical plant. Bierwirth initiated the discussion. He said that research engineers at National Distillers (through its U.S. Industrial Chemical subsidiary) had been toying with the idea of building a petro-chemical plant to produce plastics. He thought that the gas necessary for plastics production could be piped to the Midwest from the Hugoton Field by Panhandle Eastern. He convinced Maguire that Panhandle's gas should be used as the feedstock for the projected plant and that the two corporations should jointly develop the new business.

Their joint enterprise, National Petro-Chemicals Company, began operations in 1953. The plant extracted five different hydrocarbons from the gas stream: ethane, propane, butane, isobutane, and pentane. National used these hydrocarbons to manufacture several petrochemical products, including polyethylene (a polymer marketed as "Petrothene," used in packaging and insulation), industrial alcohol, ether, and ethyl chloride. The extraction of hydrocarbons reduced the BTU content of Panhandle's gas, but it was later mixed with gas from Trunkline and enhanced to meet market requirements. Panhandle then sold the gas to consumers.[13]

---

[12] George Ballard, interview by Robert B. Eckles, February 29, 1956, p. 9, PECA.

[13] "Ten Years of Progress," *Panhandle Lines*, August 1961, p. 14.

Domestic use of plastics and chemicals was rapidly increasing and National Petro-Chemical's prospects for growth seemed excellent. In its first year of operations, the company reported a net income of over $6 million.[14] The success of the polyethylene plant prompted National Distillers to construct a similar, wholly-owned facility in Houston in 1959. Upon completion of this facility in 1960, National became the second largest producer of polyethylene in the world.[15]

Before that time, however, Maguire had decided to eliminate Panhandle Eastern's operational responsibility in polyethylene. In late 1957, Panhandle exchanged its 40 percent ownership of National Petro-Chemicals for 1.5 million shares of National Distillers common stock, with a value of $35 million. Over the next thirty years, Panhandle carried this investment at its low-book value, while dividends from the petro-chemical operations contributed significant sums to Panhandle revenue. The result was a coup for the firm's stockholders. Understandably, the company attracted Wall Street's attention throughout the 1950s and 1960s because of its high return on equity. Diversification in this instance was a successful strategy which yielded a valuable block of income-producing stock without diverting management's attention from its core businesses.

Oil refining and production

Maguire had similar success in his forays into the oil industry. During 1955, Panhandle Eastern contracted an industrial engineering firm to assist in streamlining the firm's organization. Maguire wanted help in restructuring to minimize the effects of FPC rate regulation. One of the consultants suggestions was that Panhandle refine and sell the petroleum it located in its gas wells. To accomplish this, they also recommended the acquisition of a small refinery, Century Refining Company, located in Garden City, Kansas.[16] Panhandle acquired the refinery in 1957 for $800,000. Century could refine up to 4,000 barrels a

[14] Panhandle Eastern, *Annual Report, 1956*, p. 17.
[15] Panhandle Eastern, *Annual Report, 1959*, p. 17.
[16] W. C. Marris, telephone interview by Clarance M. Smith, July 22, 1993.

day, twice Panhandle's modest daily oil production of 2,000 barrels from fifty-six wells.[17]

At the same time, the company acquired all the assets of the Shallow Water Refining Company in Scott City, Kansas, and merged them into Century Refining. These facilities included local crude oil gathering lines and a refinery that produced gasoline, diesel fuel, jet fuel, asphalt, and other petroleum products.[18] This refinery was to provide a processing outlet for the firm's natural gas liquids, but substantial repairs were necessary before profitable operations could be achieved. Late in 1960, Century completed a modernization program which included the installation of a catalytic cracking unit, a propane deasphalting plant, and an alkylation plant. These improvements allowed Century to produce a better quality gasoline, diesel fuel, and asphalt and to record its first year of profitable operations in 1961.[19]

Meanwhile, Panhandle Eastern had ventured further into oil production. Its first program specifically designated for the exploration and development of new producing properties had begun in the region near St. Clair, Michigan during 1953. Subsequently, Panhandle found two fields in the southeastern portion of the state. In October 1957 the first oil discovery occurred northeast of Detroit, where Panhandle owned leases on 98,000 acres. After developing this property, Panhandle sold it (1963) to Consumers Power Company, one of Panhandle's largest utility distributor customers. Included in the sale were ten producing oil and gas fields in Macomb, St. Clair, and Missaukee counties, one underground storage field, a gas conditioning

---

[17] Century Refining, *Minutes*, February 27, 1957; and "Panhandle Subsidiary Purchases Refinery," *Panhandle Lines*, April 1957, p. 14. The original directors of Century included Robert Morgenthau, Les Fournier, William C. Marris, Fred Robinson, Frank Mathias, and Martin Kirk Hager. Hager, who had been with the company since 1936, was appointed the first president of the company, and Marris the vice-president.

[18] This conversion began in the early 1960s with the acquisition of Panhandle Eastern's Liberal gasoline plant. Century Refining later sold its oil refinery operations and became a hydrocarbon liquids marketing group. Century Refining, *Minutes*, March 8, 1957. David Hughes, interview by Clarance M. Smith, November 21, 1991, PECA.

[19] A profit of $462,000 was posted that year. "Pipeline Profile: Don Curran," *Panhandle Lines*, August 1963, pp. 12–13.

plant, and about 650,000 acres of unoperated leaseholds scattered throughout thirty-seven counties of the lower Michigan peninsula.[20]

The company's exploration in Kansas was also fruitful. In 1955, a test well drilled in Morton County, Kansas, was successful. This well, Watson 01, later produced more than 100 barrels a day from a depth of 4,500 feet and was the company's first wholly owned oil well. During the next three years, Panhandle drilled fifty-two oil wells in Kansas, thirty-three in Oklahoma, and six in Michigan. The estimated value of its oil properties jumped from $250,000 in 1956 to more than $7,000,000 by the end of 1958. Proved oil reserve estimates rose to more than 10 million barrels by year end 1958.[21]

As company-owned oil production increased, Panhandle contracted for transportation services through a 242-mile pipeline extending from Meade, Kansas to Wichita. Jayhawk Pipeline, a twelve-inch line completed during the summer of 1958, was designed to carry 45,000 barrels of oil daily to two refineries located near Wichita. The pipeline tripled the business's oil sales.[22] Don Jackson, manager of Drilling and Production, commented: "Up until now, there's been little reason to drill for oil. But as more and better marketing facilities become available in the area, this situation will change. There haven't been enough wells drilled yet to even begin to test the acreage involved. But we know there's a lot more oil there. Eventually we'll find . . . it."[23]

Anadarko Production Company

The success of the company's exploration program eventually led management to organize a new subsidiary company to drill its wells. In most respects, the Anadarko Production Company (APC) was modeled on Panhandle's 1948 spin-off, the Hugoton Production Company. Both were intended to lessen or eliminate FPC rate regulation and regulatory delay by separating gas production from gas transmission. In the case of Hugoton, the gas was dedicated to intrastate service,

[20] "Michigan Production is Sold," *Panhandle Lines*, January 1963, p. 10.

[21] "The Millionth Barrel," *Panhandle Lines*, April 1959, p. 9.

[22] Ibid., p. 10.

[23] Ibid.

thereby exempting it from federal regulation. Anadarko's production was for interstate commerce, but it could be priced higher than the gas the company already owned. Because the judicial system rejected the "fair field" pricing method, Panhandle's gas production could be priced at a competitive level only if it was transferred to a distinct corporate organization – like Anadarko.[24]

In response to the adverse ruling on the "fair field" pricing formula, Panhandle Eastern transferred all of its undeveloped acreage to Anadarko in 1959; Panhandle's 544 producing gas wells could not be transferred without an FPC abandonment certificate. The pipeline firm valued the properties located in Kansas, Oklahoma, and Texas at approximately $4 million and exchanged them for $500,000 cash and all of the shares of Anadarko common stock. APC acted quickly to develop certain Anadarko Basin acreage – due to "expiring leases and the need for drilling offset wells on such acreage," – and entered into a twenty-year contract with Pioneer Natural Gas Company. A natural gas producer and distributor headquartered in Amarillo, Texas, Pioneer served a large number of communities in the Texas Panhandle. The gas sold to Pioneer was produced in Potter, Carson, Hutchinson, and Moore Counties, Texas, and was thus exempt from FPC jurisdiction.

Anadarko's first year of full operation was 1960, and its drilling success rate was remarkable. During that year, Anadarko drilled forty-eight wells, thirty-eight of which were producers. In 1961, Anadarko constructed an eighty-four-mile intrastate pipeline from near Harper County to El Dorado, Kansas, a line that provided the Wichita–El Dorado area with its first reliable supply. By the end of 1961, Anadarko's executives were contemplating additional sales and the construction of new facilities in order to serve other Kansas industries. During its first several

[24] Tom Dufield, who began working for Panhandle Eastern in the 1950s as a young engineer, recalled that he and several coworkers were assigned to make maps of some of the early wells and company-owned production property in southwestern Kansas. The supervisor told his charges to place the letter "A" in several of the sections. Only after several years had passed were they finally able to understand the reason for the request. The marked properties in Oklahoma, Kansas, and Texas were eventually spun off into Anadarko Production Company. According to Dufield, these properties originally had been part of the Hugoton Production Company.

years of operation, Anadarko sold no natural gas to Panhandle because the firm's managers could not be certain that the FPC would allow Panhandle to pay its own subsidiary the same gas price it paid nonaffiliated companies.[25]

Panhandle soon expanded its production operations when in 1965 it acquired Ambassador Oil Company, an independent firm located in Fort Worth, for $12 million subject to production payments of $24 million.[26] Ambassador broadened Anadarko's geographic base and significantly increased its oil reserves as well as its public relations potential. The latter was a function of its newly acquired investors. The primary owners of Ambassador were members of the F. Kirk Johnson family of Fort Worth, Texas. Several movie personalities, including Jimmy Stewart and Alfred Hitchcock, were direct partners in oil and gas properties operated by Ambassador. By this time, Hollywood actors and producers had very large salaries and their financial advisors were looking for tax write-off investments for them. The financial interest these celebrities had in Ambassador entitled them to tax write-offs available to direct owners of producing properties. Neither the movie stars nor Panhandle's executives would have reason to be disappointed with Anadarko's subsequent performance.

National Helium

Maguire also brought Panhandle into a helium venture. During a series of informal discussions in 1958 and 1959, Bierwirth of National Distillers persuaded Maguire to form a partnership to construct a helium plant adjacent to Panhandle Eastern's

[25] Richard L. O'Shields, interview by Christopher J. Castaneda, June 11, 1995. PECA.

[26] Panhandle Eastern, *Annual Report, 1965*, p. 3. Anadarko's early sustained growth came under the leadership of R. B. Harkins and Richard (Dick) L. O'Shields. O'Shields joined the Panhandle organization in 1960. Prior to that, he had been a senior petroleum engineer for several different oil producers. A mechanical engineering graduate of Oklahoma University, O'Shields obtained a master's degree in petroleum engineering from Louisiana State University. O'Shields subsequently rose to the positions of vice-president (1962) and president (1966) of Anadarko. Richard L. O'Shields, interview by Clarance M. Smith, April 8, 1992, PECA.

compressor station near Liberal, Kansas.[27] Maguire was apprehensive about the partnership because it was based on a contract with the government about which he had grown increasingly wary. But Bierwirth and several of Maguire's lieutenants favored the project; Bierwirth in particular pointed out that the National Petrochemical joint venture had been successful and the National Helium project would be as well.

The federal government was concerned at this time that it would not have an adequate supply of the gas. Following a study by the Department of Defense, the Atomic Energy Commission, the Department of the Interior, and the National Aeronautics and Space Administration (NASA), Secretary of the Interior Fred A. Seaton informed President Eisenhower of the need to conserve helium sources, including natural gas. In August 1958, Seaton sent to Congress draft legislation for helium conservation. The legislature enacted the resulting Helium Act Amendments of 1960, which Eisenhower signed into law on September 13, 1960. The following year, Congress authorized an annual expenditure of up to $47.5 million for the purchase of helium from private companies. The Act provided for a twenty-two-year program to store 52 Bcf of helium at the Cliffside storage field in the Texas Panhandle.[28]

Since geologists estimated that the Hugoton and West Panhandle gas fields contained over 90 percent of the nation's helium supply, Panhandle Eastern was well-positioned to take advantage of this new policy. The firm controlled more than 2,500 of the gas wells in those fields. Working with National Distillers, Panhandle built one of the five helium extraction plants constructed as a result of the new law. The Helium Conservation Pipeline connected all of these plants, as well as four other private ones (Map 7.1).

The U.S. space program, led by NASA, was the single most important market for helium. Large quantities were used in the Atlas and Titan missiles that sent the Gemini craft into outer space. Helium pressurized the fuel compartments in rockets and missiles and prevented the thin skin of missile housings from collapsing. The gas also proved to be a valuable com-

[27] For a readable history of helium utilization in the twentieth century see, Clifford W. Seibel, *Helium: Child of the Sun* (Lawrence: The University Press of Kansas, 1968).

[28] Seibel, *Helium*, p. 119 and Panhandle Eastern, *Annual Report, 1961*, p. 26.

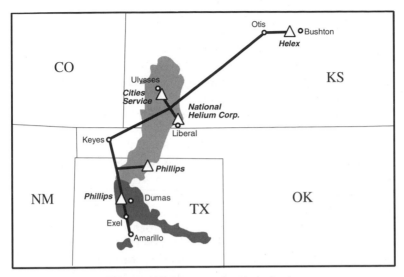

Map 7.1 Helium production plants

ponent in nuclear reactors because it absorbed heat and had a high resistance to radioactivity. It had other high technology applications in computer development, superconductivity research, and medicine.

To capitalize on this new market, Panhandle Eastern and National Distillers – working as equal partners – organized the National Helium Corporation in June 1960.[29] For Panhandle Eastern, F. J. McElhatton (National Helium's first president) and engineer Ross Wilson participated in negotiations with the Department of Interior for the helium sales contract. In October 1960, National Helium signed a twenty-two-year contract with Interior and began building a $30 million plant designed to process 850 MMcf/d of natural gas from Panhan-

[29] Maguire named F. J. McElhatton, vice-president of gathering and transmission of Panhandle Eastern, as the first president of the new corporation. The vice-president and manager of National Helium was Ross W. Wilson who later became president as well. Wilson began his career in 1946 as an employee of the U.S. Bureau of Mines at its Amarillo helium extraction plant. He left government service in 1952 to join Panhandle Eastern. His first posting was Tuscola, Illinois where he was Panhandle Eastern's representative in the initial operating years of National Petro-Chemicals Corporation.

dle. Although the helium accounted for only .4 percent of the total natural gas stream to flow through the plant, 90 percent of the helium was recovered.[30] The new subsidiary was authorized to sell its helium for $11.78 per Mcf, up to a total of $15.2 million per year.[31] The plant extracted helium from the gas stream by passing it through two ten-story cold boxes capable of reaching a temperature of $-300°C$. At 500 lbs/psi and -270 degrees, methane, the principal ingredient of natural gas, liquefies, but helium and nitrogen remain gaseous, allowing for their separation from the liquified methane. The gaseous helium and nitrogen mixture was separated and later fractionated to a helium rich mixture that was then piped to the underground storage field. Panhandle's natural gas returned to the firm's transmission system.[32]

Over 400 persons attended the dedication for the helium facility on September 24, 1963. Maguire and Panhandle attorney Raymond Shibley traveled by private rail car to Dodge City, Kansas, eighty miles from the site at Liberal. There they got into a black stretch limousine, with a second identical vehicle following in case the first one broke down.[33] Dignitaries in attendance included John Anderson, Jr., Governor of Kansas; Henry Wheeler, the deputy director of the Bureau of Mines; and Robert Dole, congressman from Kansas. Dole announced that the flag being flown over the Capitol in Washington that day would be presented to National Helium.[34]

The helium venture began on a happy note. At the dedication, Charles Zimmerman, NASA's director of aeronautical research, gave the keynote address. Helium, he said, was vital to a U.S. space program locked in competition with the USSR.

---

[30] Harry C. Barnes, *From Molasses to the Moon: The Story of U.S. Industrial Chemicals Co.* (New York: U.S. Industrial Chemicals Company, 1975), 130. The contract between M. W. Kellogg and Panhandle Eastern was dated April 16, 1962.

[31] "World's Largest Helium Plant," *Panhandle Lines*, November–December, 1961, p. 5.

[32] Eugene Guccione, "New Approach to Recovery of Helium from Natural Gas," *Chemical Engineering*, September 30, 1963. Gene T. Kinney, "World's Biggest Helium Plant Opens," *Oil and Gas Journal*, September 30, 1968, pp. 54–60.

[33] Raymond M. Shibley, interview by Christopher J. Castaneda and Clarance M. Smith, January 13, 1994, PECA.

[34] Ibid.

Figure 7.1 National Helium Corporation (cryogenic section)

"Exploration of space," he said, "will go forward, if not by us, then by others." If America's space program developed rapidly, the U.S. might reach the moon before Russia.[35] He also praised the "government–industry partnership that is making this massive conservation effort possible without using taxpayer

[35] "Dedication Day," *Panhandle Lines*, vol. 21, no. 3, (October 1963), p. 3.

Figure 7.2 Robert Dole and William G. Maguire at National Helium
dedication

funds. Private industry has invested about $100 million in five
new plants to assure our nation the helium supplies needed in
the future."[36]

Panhandle's executives were for a time as pleased with this
undertaking as was NASA.[37] It appeared initially that the company would profit immensely from National Helium. In addition to supplying the facility's total natural gas needs, Panhandle Eastern received profits from the subsidiary's helium sales
to the government for more than $11 per Mcf. A law suit filed
by disgruntled royalty owners which Panhandle lost eventually erased the venture's profits.[38] The promising operation
suffered another blow when in 1973 the government stopped
taking deliveries of helium, well before the contract termination date. Panhandle Eastern and National Helium subsequently were forced to sue to recover their investment and to
transform the helium facility into a natural gas liquids extraction plant.[39]

[36] Ibid. Also see, "Negotiate a Helium Plant," *Panhandle Lines*, March 1961, 2.
There were originally fourteen applications from which Secretary of the
Interior Stewart L. Udall chose four on March 21, 1961.
[37] "Dedication Day," *Panhandle Lines*, vol. 21, no. 3 (October 1963), p. 5.
[38] Panhandle paid approximately $77 million to royalty owners. Panhandle
Eastern, *Annual Report, 1984*, p. 48 and Panhandle Eastern, *Annual Report,
1985*, p. 51.
[39] Richard L. O'Shields, interview by Christopher J. Castaneda, June 11, 1995.
PECA. The National Helium Company resumed service in 1989.

The crisis of transition

On March 1, 1965, when seventy-nine-year-old William Maguire marked his thirty-fifth year at Panhandle Eastern at a special luncheon, he had much to celebrate. The company's rather conservative diversification enterprises had for the most part been successful. Maguire's intensely personal, financially oriented style of management appeared to have brought Panhandle through the problems stemming from the Phillips decision. But soon the firm faced a new crisis. During the late fall of 1965, Maguire became ill and died on September 28.[40]

One week after his death, the Panhandle board elected W. K. Sanders, sixty, president of Trunkline Gas, to be president and CEO of Panhandle Eastern and Trunkline.[41] Sanders was a thirty-seven-year veteran who had been with the enterprise since 1928. This was a compromise decision – a holding action stemming from the fact that Maguire had declined to designate an heir. There were rumored heirs including Fred Robinson, Maguire's son-in-law; Bill Marris, VP Rates and Regulatory Affairs; and King Sanders, president of Trunkline. But Maguire had never made his intentions clear, and the board had to reach its own decision, under pressure to act quickly. They opted for Sander's longevity with the organization, his commonsense approach to management, and his practical knowledge of pipeline operations.[42]

King Sanders quickly asserted his authority. He had worked his way up from pipeline crew member in the early days of Panhandle Eastern, and he now introduced his own style of hands-on management. He enjoyed knowing employees at all levels of the company, and he soon decided to relocate all of those executives who were in New York to Houston. Houston was the location of Trunkline's headquarters. It was a city identified with oil and gas production rather than the Wall Street financial community. It was also the location of Sanders's home.

---

[40] "William G. Maguire . . . 1886–1965," *Panhandle Lines*, November 1965, p. 3.

[41] Panhandle Eastern, *Annual Report, 1967*, p. 3. "Pipeline Profile," *Panhandle Lines*, May 1958, p. 2.

[42] Several interviews deal with the Maguire–Sanders transition, including those of Bertha Sanders, William C. Marris, and J. M. "Mack" Price, PECA.

Figure 7.3 After Maguire: Fred Robinson, W. K. Sanders, and William C. Keefe

As for operations, Sanders said there would be some changes, but most would be minor. "For one thing," Sanders said, "we're going to put more emphasis on long-range planning, not only with respect to expansion but on all other phases of the business. This will help not only our own people but also our customers." In regard to the diversification strategy, Sanders said that "since the early 1950s, Panhandle has recognized the wisdom of diversification. The National Distillers–USI plant at Tuscola, National Helium Corporation and the recent Ambassador acquisition by Anadarko are good examples." Like Maguire, he wanted to keep Panhandle's investments in areas closely related to natural gas.[43]

On other matters, however, Sanders brought in a new approach to management.[44] He set out to change the rigid hierar-

---

[43] "Mr. Sanders Answers Questions," *Trunkline*, December 1965, p. 5.

[44] At one point, Sanders discussed the rumors of a merger of Panhandle Eastern and Trunkline: "I am sure there have been a lot of rumors about this, and gossip of this type travels via the grapevine faster than it would if we broadcast it on our microwave system. Panhandle and Trunkline actually have a different set of operating problems and one company isn't going to swallow up the other by merger or otherwise. I do think, however, that you can expect

chies Maguire had created and gave the younger members of Panhandle more decision-making power. As Sanders noted, the "second level of management was highly concentrated. Our internal channels were restricted, with too many matters funneling upward before decisions could be made. I firmly believe that many more decisions should be made at lower management levels. We gave this matter considerable study and have made the necessary changes."[45] He also allowed more upward mobility among managers.

Regarding Panhandle Eastern, Trunkline, and Anadarko, Sanders noted: "These groups are now staffed with several new vice presidents, each of whom was drawn from within our organization. Along with vigor, they have experience in their areas of responsibility, providing the management strength and competence needed. . . . Each pipeline system now has an executive vice-president responsible for all matters and reporting to our president. This, in turn, has enabled us to assign corporate planning and policy formation responsibilities to officers who have been freed of daily operating matters."[46] The new organization embraced the profit center principle, with a high degree of decentralization.[47] The board of directors as well as the management style changed during this era. Maguire's hand-picked board members were slowly beginning to be replaced. Jack Bierwirth, National Distillers chairman and a long-time associate of Maguire's, retired in 1970 after thirty years of service on the Panhandle Eastern board. Drummond C. Bell, President of National Distillers, replaced Bierwirth.[48] A 1969 retiree from the Trunkline board was Lewis W. MacNaughton, a renowned geologist and a partner in a Dallas-based engineering and consulting firm with Everett DeGolyer. Charles E. Main, another board member who had served with Maguire since 1943, also retired in late 1970.[49]

With new leadership and a new structure, Panhandle Eastern appeared to be in an excellent position to cope with the

a greater cooperation and coordination between the two companies." "Mr. Sanders Answers Questions," *Trunkline*, December 1965, p. 5.

[45] "Panhandle Eastern Reports," *Panhandle Magazine*, Spring 1969, pp. 15–16.

[46] Ibid.

[47] Luke Scheer, "Epilogue: On the 1980s," p. 5, PECA.

[48] Panhandle Eastern, *Annual Report, 1969*, p. 3.

[49] Panhandle Eastern, *Second Quarterly Report to Stockholders, 1969*.

problems the natural gas industry was experiencing under the regulatory regime of the postwar years. New strategies would certainly be needed if the company was going to be as successful in the next two decades as it had been under Maguire's forceful hand. There was no reason to believe that CEO Sanders and his successor – as of 1969 – would not be up to that challenge.

PART III

Search for a new equilibrium

8

Strategic innovations

"Our problem," said a gas company spokesman during the harsh winter of 1976–7, "is that . . . [customers are] pulling the gas out of the system as fast as we can put it in. We have a system built for four degrees below zero and it's 11 below outside."[1] Gas shortages forced schools to close, factories to lay off workers, and homeowners to suffer through frigid temperatures without heat. There was simply too little natural gas available in the interstate system to fuel the nation through that terrible winter.[2]

The legacy of the Phillips Decision was a natural gas indus-

[1] Steven Rattner, "Gas Crisis has Complicated Origins," *New York Times*, January 30, 1977.

[2] For a discussion of the problems inherent in regulated industries see Stephen Breyer, *Regulation and its Reform* (Cambridge, Mass.: Harvard University Press, 1982) and Paul W. MacAvoy, *The Regulated Industries and the Economy* (New York: W. W. Norton, 1979). Also see Kahn, *The Economics of Regulation*, 2 vols. Kahn's work on natural gas pricing was instrumental in the development of the two-tiered pricing system. For an evaluation of Kahn's work see Thomas K. McCraw, *Prophets of Regulation*, pp. 235 and 222–99. Also see FPC, *A Staff Report on National Gas Supply and Demand* (Washington: FPC, Bureau of National Gas, September 1969); Panhandle Eastern Pipe Line Company Application to the New York Stock Exchange, Inc. (April 27, 1977), pp. 34–5, PECA; and Steven Rattner, "Cold Deepens Natural Gas Shortage; Continued Supply Cutbacks Expected," *New York Times*, January 19, 1977.

Table 8.1. *Natural gas supply in the United States,*
1958–1973 (Tcf)

Year	Production	Reserve additions	Reserves/ production ratio	Total gas wells
1958	11.4	18.9	22.1	5,005
1959	12.4	20.6	21.1	4,931
1960	13.0	13.9	20.1	5,129
1961	13.5	17.2	19.9	5,459
1962	13.6	19.5	20.0	5,353
1963	14.5	18.2	19.0	4,570
1964	15.3	20.3	18.3	4,694
1965	16.3	21.3	17.6	4,482
1966	17.5	20.2	16.5	4,321
1967	18.4	21.8	15.9	3,659
1968	19.4	13.7	14.8	3,456
1969	20.7	8.4	13.3	4,083
1970	22.0	37.2[a]	13.2	3,840
1971	22.1	9.8	12.6	3,830
1972	22.5	9.6	11.8	4,928
1973	22.6	6.8	11.1	6,385

[a]Represents addition of Alaskan reserves.
Source: American Gas Association, *Gas Facts*, 1958–73. Also see American
Association of Petroleum Geologists and American Petroleum Institute (from
tables in House of Representatives, Committee on Interstate and Foreign Commerce, Hearings on Natural Gas Supplies, 94th Cong., 1st sess., June, July,
1975, pp. 1043–51); and Sanders, *The Regulation of Natural Gas*, p. 126.

try unable to adjust quickly to such a wide spread shortage.
Wellhead price controls had discouraged exploration and gradually cut down the gas reserves available for interstate sale by
pipelines to customers. As Stephen Breyer and Paul MacAvoy
pointed out, the failure to develop new reserves early in the
1960s was already curtailing production by the end of the decade.[3] In 1968, as Table 8.1 indicates, annual gas production for
the first time exceeded the discovery of new reserves, and between 1968 and 1975, new discoveries yielded only 45 percent
of total domestic consumption.[4] By the 1970s, the industry was

[3] Stephen G. Breyer and Paul W. MacAvoy, *Energy Regulation by the Federal
Power Commission* (Washington, D. C.: The Brookings Institution, 1974), p. 73.
[4] See William D. Smith, "Natural Gas: Short, and Getting Shorter," *New York
Times*, September 28, 1975, sec. II, p. 1.

Table 8.2. *Curtailments by U.S.*
interstate pipelines, 1970–1977

Year	Curtailed deliveries (bcf)
1970	17
1971	286
1972	649
1972–73	1,031
1973–74	1,191
1974–75	2,418
1975–76	2,975
1976–77	3,790

Source: FPC, *Natural Gas Survey* (Washington, D.C., 1975), vol. 1, p. 105; and FPC, *Annual Report*, p. 16.

ripe for a crisis, and the combined effects of the 1973 Arab oil embargo and the extremely cold winter of 1976–7 left companies like Panhandle unable to provide the nation's consumers or producers with the energy supplies they needed (Table 8.2).[5]

Neither Panhandle's executives nor the industry's regulators were sitting on their hands waiting for this crisis to develop, and the interaction between the political system and the firm is the central theme of the pipeline's history in the 1970s.[6] The government attempted to encourage exploration and production for interstate commerce; it also subsidized the develop-

[5] Richard Vietor argued that "The concrescence of these three trends – depletion, concentration, and environmentalism – nearly overshadowed a polity already shaken by a broader crisis of a confidence that stemmed from the Vietnam War." Richard H. K. Vietor, *Energy Policy in America since 1945*, p. 194. Also see Steven Rattner, "Gas Crisis Has Complicated Origins," *New York Times*, January 30, 1977; Edward Cowan, "Natural Gas Allocation Dilemma," *New York Times*, February 23, 1975; and Edward Cowan, "Northeast Warned It Faces Winter Natural Gas Crisis," *New York Times*, September 15, 1974.

[6] In David Howard Davis, *Energy Politics* (New York: St. Martin's Press, 1978), Davis suggests that the FPC deserves blame more than other agency, or the judicial system, for its regulatory policies which weakened the structure of the natural gas industry (p. 131). Clearly, the FPC's powers were shaped and created by other agencies and the judicial system in particular. Thus, it is neither particularly productive nor instructive to simply blame the FPC; it is better to examine the process through which the FPC's powers were instilled.

Table 8.3. *Panhandle Eastern Pipe Line Company: Earnings and financial data, 1968–1977*

Year	Operating revenue (millions)	Operating income (millions)	Net income (millions)
1968	337	59	42
1969	380	71	44
1970	419	77	50
1971	451	77	46
1972	500	89	89
1973	519	100	64
1974	564	117	69
1975	659	124	72
1976	825	132	88
1977	1,204	182	101

Source: Panhandle Eastern Pipe Line Company, *Annual Report*, 1973, 1976, 1982.

ment of alternative energy sources. These efforts were frequently handicapped by an environmental movement which had achieved considerable support and political power by the 1970s. But Panhandle and other firms in natural gas were still able to take advantage of the subsidies, increase earnings, and in some cases actually help alleviate the shortages created by regulatory failure. For Panhandle these were particularly trying times because the company's management was still operating in the formidable shadow of the late William Maguire. As the company's executives struggled to deal with a trying political context, they were simultaneously attempting to put their own stamp on the business's structure, operations, and leadership (Table 8.3). As they well knew, many successful companies had found the transitions from a powerful founder to a new cadre of professionals extremely difficult to make. These were, then, testing years for both the company and the regulatory system.

The synthetic gas experiment

As early as 1968, Panhandle Eastern had begun to deal systematically with the long-term gas supply problems created by price controls. In that year, CEO King Sanders formed a corporate planning group that he charged with the responsibility for acquiring adequate supplies of gas to serve existing and future demand. The primary leader of this effort through the 1970s was George Kurk, senior vice-president. He had worked for Peat, Marwick, Mitchell & Company for ten years prior to joining the Panhandle organization in 1958, and had served as controller and vice-president of Anadarko during the 1960s.[7] As of 1970, Kurk and his group worked closely with Panhandle's new CEO and president, Richard "Dick" O'Shields.[8] O'Shields represented the new breed of executive in the company. Only forty-two years old, he had made a name for himself as an engineer in the employ of the Sid Richardson organization in Fort Worth, Texas.[9] His strong engineering background and previous experience at Panhandle as president of Anadarko Production Company prepared him to take the Panhandle organization in new directions, including a program to utilize coal directly and to produce synthetic coal gas.[10]

[7] Biographical data sheet on George J. Kurk dated September 8, 1983, PECA. Jack Schomaker, interview by Clarance M. Smith, April 22, 1992, PECA.

[8] "Move Ahead in Pipeline Firm," *The Kansas City Times*, January 4, 1968. *Panhandle Magazine*, Autumn 1971, p. 11. In the late 1960s, Panhandle Eastern's management changed rapidly. The reign of King Sanders was fairly short-lived. In 1968, William C. Keefe became Panhandle's new president, a promotion made possible by the election of King Sanders, who remained CEO, as chairman of the board following the retirement of Fred Robinson. In 1970, Sanders retired at the mandatory age of sixty-five and Bill Keefe was elected chairman of the board. Dick O'Shields, who held the Panhandle Eastern executive vice-president position created in 1968, succeeded Bill Keefe as president and was named CEO as well. During his brief reign, Keefe argued that Panhandle was still unable to obtain a reasonable rate of return from the FPC, and he emphasized the desirability for diversification while blaming the FPC for Panhandle's lower profit margins. Keefe was married to Maguire's niece, Barbara. See William C. Keefe, Speech to Shareholders, 1969, PECA.

[9] "Profile: Richard L. O'Shields," *The Oil and Gas Journal*, September 14, 1970, pp. 143–4.

[10] Panhandle also planned to produce natural gas in the undeveloped Powder River Basin of Wyoming. In 1970, Panhandle embarked on its first effort to expand its geographical supply base by purchasing casinghead gas from producing oil wells in the Powder River Basin of Wyoming. The gas was delivered to Kansas–Nebraska Gas Company, Inc. under a displacement and

In a sense, the synfuels plan represented a return to the days when Panhandle was first pushing natural gas as a replacement for the manufactured product. But in the 1970s, the technology was far more advanced than it had been in the 1920s, as was the regulatory system. During the 1970s, coal which comprised 90 percent of U.S. energy reserves, seemed to be a very promising fuel. The domestic supply was enormous and was free from foreign political and economic entanglements. The U.S. Geological Survey assessed the amount of recoverable coal reserves in the United States at between 150 and 438 billion tons, an amount equivalent to a 200-year supply. With coal, too, there would be no twelve- to fourteen-year waiting period, as was the case with nuclear plants.[11]

Since World War II, coal's contribution to total energy use had declined substantially, due in large measure to the increase in natural gas utilization. After the Arab oil embargo and America's realization of its dependence on foreign oil, however, the country's coal reserves were literally rediscovered. One writer described coal as "a somewhat frumpy middle-aged ballerina rushed out of semiretirement to fill an unanticipated gap in a show that must go on. Suddenly, the old girl is back in demand." In this case, the "old girl" received applause from the federal government and a fierce round of boos from environmental groups.[12]

Panhandle managers agreed with the government and they approached coal from two perspectives: they planned to use it to make synthetic gas; and they decided also to get into coal

exchange agreement in which Kansas–Nebraska made volumes available to Panhandle from sources of production in Texas and Oklahoma. Another program initiated in 1970 involved federally approved advance payments for supplies being developed in the Denver–Julesburg Basin in Colorado. Again, through a displacement and exchange agreement, Colorado Interstate Gas Company provided volumes to Panhandle Eastern in Kansas. Both efforts were stopgaps. Panhandle management continued searching outside the continental United States for large volumes of long-term supply. "The Promise of the Rockies," *Panhandle Magazine*, no. 2, 1979, p. 7.

[11] Wheeler, Romney. "The Long, Hard Road for Coal," *Panhandle Magazine*, no. 1, 1978, p.1.

[12] Mel Horwitch, "Coal: Constrained Abundance," in *Energy Future,* Robert Stobaugh and Daniel Yergin, eds. (New York: Random House, 1979), pp. 79–80. Also, see Edmund Faltermayer, "Clearing the Way for the New Age of Coal," *Fortune*, May 1974, pp. 215–38.

production. The manufacture of coal gas was certainly not a new process, but the modern technology differed substantially from the earlier process, largely as a result of advances made in Europe and South Africa. The newer techniques could produce nearly pure methane, the principal ingredient of natural gas, and this product could be used in residential, commercial, and industrial applications.

Panhandle Eastern launched its synthetic coal gas project, WyCoalGas – a joint-venture with Peabody Coal Company – in the early 1970s.[13] Peabody was one of the nation's largest coal producers. It owned reserves totaling about 165 million tons in Illinois and approximately 500 million tons in Wyoming. These extensive reserves were dedicated to Panhandle Eastern's gasification project, which also received water rights and surface lands for its Wyoming plant site. A large water supply was essential to gasification, and in semiarid Wyoming, water was scarce and costly. Once an adequate water supply was assured, the plant would be able to produce 250 Mcf/d a year.[14] Panhandle hired M.W. Kellogg Company and American Lurgi as consultants to assess the technological and ecological aspects of this project. WyCoalGas also contracted with the South African

[13] Jack Schomaker provided a detailed account of the origin and plans for WyCoalGas. Other pipeline firms considered the possibility of engaging in coal gasification projects as early as 1973. The first modern project was designed by El Paso Natural Gas in conjunction with Consolidation Coal, a subsidiary of Continental Oil; see Vietor, *Energy Policy in America since 1945*, p. 293. Pacific Lighting Corporation and Texas Eastern Transmission Corporation, in a joint venture called Wesco, also planned to build a plant on Navajo lands in New Mexico. The Wesco project was the first coal gasification project to receive full FPC consideration. The owners promoted the plan for six years, but it was finally scrapped in 1979 after difficulties in obtaining financing and disagreements with the Navajo Tribal Council. Castaneda and Pratt, *From Texas to the East*, pp. 213–16; and Vietor, *Energy Policy in America since 1945*, p. 294. Also see, Federal Energy Administration, *Final Task Force Report on Coal Project Independence* (Washington, D.C.: GPO, November 1974); James K. Harlan, *Starting with Synfuels* (Cambridge, Mass.: Ballinger, 1982); and Ernest J. Yanarella and William C. Green, eds., *The Unfullfilled Promise of Synthetic Fuels: Technological Failure, Policy Immobilization, or Commercial Illusion* (New York: Greenwood Press, 1987).

[14] Panhandle Eastern, *Annual Report, 1970*, p. 16; Panhandle Eastern 1977 proxy statement, p. 38; and Panhandle Eastern, *Annual Report, 1972*, p. 15.

government-owned oil firm, SASOL, for design assistance and advice. SASOL's African plant had for many years successfully manufactured a low-Btu gas using the Lurgi process – a German technology that had been developed in World War II.

The new technology failed to persuade the environmental opponents of the project that WyCoalGas should be built and operated. Nor were they convinced by the Environmental Impact Statement or by Panhandle Eastern's waste water recycling program. Solid wastes were to be returned to the coal mine for disposal in accordance with the Resource Conservation Reclamation Act and the requirements of the Water and Land Divisions of the Wyoming Department of Environmental Quality. The Air Quality Division of Wyoming assumed responsibility for oversight of the airborne emissions. Panhandle's managers had faced environmental issues before, but their experience with pipeline activities had not prepared them for the fierce battles they had to fight over this plant.[15]

These struggles delayed plant construction and became a full-time business function for WyCoalGas. Several Panhandle employees moved to Wyoming to counter the charges of the environmental groups that gasification would be detrimental to Wyoming's air, water, and wildlife.[16] In an effort to develop an amicable relationship with the state government, the company produced and donated to the state a thirty-minute Wyoming promotional film. Television personality Curt Gowdy, a Wyoming native, narrated the film. But many state legislators and such activist groups as the Sierra Club remained hostile.[17]

Although Panhandle was still hopeful about gasification, the delays forced the firm to write-down its capitalized costs. In

[15] Roger Segura, interview by Clarance M. Smith and Christopher J. Castaneda, October 7, 1993, PECA. Segura served as Environmental Coordinator for the WyCoalGas project. He later became Division Manager for the Eastern Division of Panhandle Eastern Pipe Line Company.

[16] J. M. "Mack" Price led WyCoalGas' political efforts in Wyoming.

[17] J. M. "Mack" Price, interview with Clarance M. Smith and Christopher J. Castaneda, September 10, 1993, PECA. Recognizing that a Wyoming plant site was a difficult proposition, the company undertook in 1976 a study with the city of Wichita, Kansas, to determine the feasibility of a municipal gasification plant. Under the guidelines set forth in this plan, Peabody would transport the coal by rail from Wyoming to Kansas. These discussions proved unfruitful.

1976, the pipeline took a $4.5 million loss on the development of the water supply. Three years later it wrote off $18 million for unutilized coal under contract with Peabody. Panhandle and Peabody tried to use part of their reserves for other purposes. They organized the North Antelope Coal Company, which contracted with Arkansas Power & Light to supply fuel on a long-term basis. Constructed in the late 1970s and capable of producing over 5 million tons per year, North Antelope was initially a profitable venture, but Panhandle had little use for it after interest in coal gasification dissipated. Panhandle subsequently sold back to Peabody for $52 million its 50 percent interest in North Antelope.[18]

Even direct and indirect government support had been unable to get Panhandle's coal projects moving in the 1970s and 1980s. Under the Energy Security Act, which became law in June 1980, the government established the Synthetic Fuels Corporation, which could grant up to $88 billion in federal financial assistance, including loan guarantees.[19] In addition, federal regulators supported the Gas Research Institute (1978) by allowing pipelines to include the cost of many significant research and development projects in their tariff structure.[20]

At one point, the possibility of federal subsidies had encouraged Panhandle to attempt to revitalize WyCoalGas. The pipeline brought in two new partners to help cover the project's mounting costs. These companies, California's Pacific Gas & Electric, and the German utility, Ruhrgas, wanted to learn more about gasification for their own future projects, and the expanded venture successfully applied for funding from the Department of Energy to complete important elements of its design and development work.

Despite the encouragement offered by the Gas Research Insti-

---

[18] Panhandle Eastern, *Annual Report, 1988*, p. 4. Panhandle sold North Antelope to Peabody in 1989.

[19] Sabrina Willis, "The Synthetic Fuels Corporation as an Organizational Failure in Policy Mobilization," in *The Unfulfilled Promise of Synthetic Fuels*, Ernest J. Yanarella and William C. Green, eds., p. 71. Willis argues that in "The problems of the Synthetic Fuels Corporation have been so many and so serious that it is difficult to determine exactly why the corporation failed to come close to achieving the goals set for it by Congress in the Energy Security Act."

[20] Ibid., pp. 72–3.

tute and the Synthetic Fuels Corporation guarantee, Panhandle finally abandoned WyCoalGas in March 1982. Inflation and high interest rates had raised the total projected costs to more than $3.5 billion, making it no longer economically viable. With an estimated price of over $10 per Mcf – far above the price of natural gas – the synthetic product could only be sold if it received substantial federal subsidies.[21] After the election of President Ronald Reagan, however, subsidization of gas projects was abandoned. Panhandle's management saw no prospects that support would be renewed. They decided to turn away from this failed entrepreneurial venture and cut their losses.[22]

Coal production

A similar fate awaited Panhandle's move into coal mining. This undertaking was the closest the pipeline firm came to adopting the conglomerate strategy so popular during the 1960s and 1970s. Many U.S. executives became convinced during these years that they could profitably run any sort of enterprise, even one whose technology and markets were unrelated to those of their core businesses. Most were disappointed, as was Panhandle's management. In the mid-1970s, however, the prospects for coal mining looked good. Based on the recommendation of corporate planning, Panhandle acquired the Youghiogheny and Ohio (Y&O) Coal Company in 1976. The Osborne family of Cleveland, Ohio, had founded Y&O in 1902 and had remained involved with its operations since that time.[23] Y&O's mines were located in the Ohio River valley, and its assets included extensive (approximately 400 million tons) coal reserves in Ohio, West Virginia, and Pennsylvania.[24] Much of the coal had a high sulfur content (4–5 percent), but Panhandle's corporate planning department projected increased industrial demand for coal, including the high sulfur fuel.

---

[21] Tussing and Barlow, *The Natural Gas Industry*, pp. 72–3.

[22] The Energy Security Act of 1980 authorized the Synthetic Fuels Corporation to use $17 billion to support commercial synfuels projects. See Vietor, *Energy Policy in America since 1945*, pp. 324–40.

[23] "Panhandle Acquires Y&O Coal Co.," *Panhandle Lines*, no. 3, 1976, p. 9.

[24] Y&O had over 400 million tons of coal reserves in the three states. About 75 million of the total reserves were committed to the three existing mines and the Cadiz Portal mine. "Panhandle Acquires Y&O Coal Co.," *Panhandle Magazine*, no. 3, 1976, p. 9.

Figure 8.1 Y&O Coal Company, 1976

Panhandle contracted with a consulting firm, the Arthur D. Little Company, for an assessment of the marketability of high sulfur coal and the technology available to utilities for treating the resulting emissions. The report concluded that sulfur amelioration technology was rapidly improving and that stack gas sulfur removal would be a cost-effective method for treating sulfur emissions. In addition to selling coal to utilities, Panhandle initially envisioned the construction of a coal gasification plant at the site. As Chief Executive Officer O'Shields explained, "The principal objective of the Company in making this acquisition is to broaden its base over the long term. Our continuing studies of the nation's energy needs indicate a strong demand for coal over an extended period. This outlook, coupled with the sound organization and strong coal reserves position of Y&O, makes the acquisition an appropriate investment. . . ."[25]

Panhandle paid Y&O 2,197,198 shares of its common stock

[25] Remarks by Richard L. O'Shields at Annual Meeting of Shareholders of Panhandle Eastern Pipe Line Company, New York City, April 27, 1977.

and \$2.7 million in cash for all of the coal firm's outstanding shares. Before Panhandle could complete the acquisition, Y&O had to divest a small gas operation that it owned, a process that required nearly a year to complete.[26] Then, Panhandle entered this new business. The pipeline's executives retained William M. Osborne, Jr., as president and chief executive officer and sent George Broussard from Panhandle to serve as a vice-president and assistant to Osborne. Broussard quickly realized that Y&O's problems could not be solved by the pipeline's traditional methods of expending more capital and requesting increased rates. "Y&O was really kind of a watershed in the company's history," he said, "in that when they got Y&O, that was the first time that they ever got a hold of something that they couldn't solve. It was sort of a microcosm of what we maybe have seen on a national scale. There was a long time from the [World War II] period all the way up to Vietnam. Anything that we wanted to do we basically could do. We sent people to the moon. But when Vietnam came along, we began to realize that there were some things that maybe we could do but we were not prepared to accept the pain and strain of doing it and maybe we can't even do it. That was kind of the way Y&O was."[27] Certainly, Y&O created its share of "pain and strain."

Labor relations were a quagmire. Y&O operated its mines with union labor, and within a year of the acquisition, the United Mine Workers called a nationwide strike which lasted into the following year. Panhandle's managers were unprepared for the labor problems associated with a large number of

[26] Harry Welch, interview with Clarance M. Smith, April 9, 1992, PECA; and Panhandle Eastern, *Annual Report, 1976*, p. 37. One snag in the operation of the Ohio company held up Panhandle's acquisition for almost a year. Although the primary business of Y&O had always been underground coal mining, it had ventured into the retail natural gas business in the early 1930s. A small gas field had been discovered amidst Y&O's coal mining properties in West Virginia. Local inhabitants had begun tapping the gas supply for their own use, a poaching activity which eventually led to Y&O's decision to sell the gas. At the time of Panhandle's acquisition, Y&O was still producing this gas and selling it to end users, a function that technically placed Y&O in violation of the Public Utility Holding Company Act of 1935.

[27] George Broussard, interview with Clarance M. Smith and Christopher J. Castaneda, August 22, 1991, PECA.

coal miners and with problems like coal workers' pneumo-coniosis (black lung disease).

The comforting findings of the Little report that technology would soon be available to make high sulfur coal marketable turned out to be wrong. Y&O's coal was becoming increasingly difficult to market and Panhandle's options for its utilization were rapidly diminishing. Broussard recalled that Panhandle, along with most of the major oil firms, knew "the country [was] going to have a crying need of energy, and we were not going to let these little things like high sulfur . . . slow us down. . . . And that was wrong. That was just a wrong assessment."[28] The unex-pectedly high cost of sulfur removal prompted utilities to pur-chase and transport low sulfur coal located in western states.

In 1977, the National Academy of Sciences underscored the environmental problems with coal generally when it concluded that carbon dioxide emissions from fossil fuels were contribut-ing to global warming. The Academy warned that the resulting "green house" effect "might pose a severe, long-term global threat." This ominous note strengthened the environmental opposition to high-sulfur coal in particular, and as the product declined, so too did Y&O's profitability.[29]

Unable to run Y&O at a profit, Panhandle finally decided to sell the company. It had acquired the firm at a cost of nearly $60 million during an era of energy shortages and rising prices. By the mid-1980s, the shortages were over, prices were down, and coal was no longer a particularly attractive fuel. The entire U.S. coal industry was by that time in decline, and Panhandle was unable to find a buyer. Eventually, it took a $71 million write-down on the assets and in 1987 shut down the last of Y&O's mines (keeping open only administrative offices to han-dle several labor-related matters). Panhandle closed the door on conglomeration in 1989, when it sold Y&O for a $15 million after tax gain. By that time, many other large U.S. corpora-tions had learned the same lesson and had closed the door on the conglomerate strategy.

---

[28] Ibid.

[29] Horwitch, "Coal: Constrained Abundance," in *Energy Future*, p. 92. National Academy of Sciences, *Energy and Climate: Outer Limits to Growth* (Washing-ton, D.C.: NAS Geophysics Board, 1977). George M. Woodwell, "The Carbon Dioxide Question," *Scientific American*, January 1978, pp. 34–43.

Offshore contract drilling

More successful in financial terms was Panhandle's move into contract drilling. In 1972, the company formed a joint venture with Dixilyn Drilling Company. M. O. "Mac" Boring, Jr., of Midland, Texas, had founded Dixilyn in 1945 as a contract onshore drilling firm. Eventually he had shifted Dixilyn's focus to offshore drilling. By 1972, Dixilyn management was ready to tackle the high stakes business of North Sea exploration, but the company first needed an equity partner.[30] Panhandle Eastern decided to provide capital to a joint venture named Pel–Lyn. The two partners were subsequently joined by a third, K/S Godager International A/S of Oslo, Norway. Godager's inclusion spread the capital burden, provided expertise in the North Sea, and made it more likely that the Norwegian government would view their undertakings with favor. The North Sea was divided into five sectors, the largest of which were controlled by the United Kingdom and Norway, and the Norwegian government placed severe restrictions on firms that did business in its sector. Godager certainly enhanced the likelihood of success should the partnership ever bid on jobs in that area.[31] Panhandle and Godager contributed $10 million and $14.2 million, respectively. Dixilyn contributed its U.S. drilling equipment, which had a book value of $11 million. Pel–Lyn Godager started by constructing two large semisubmersible drilling rigs, stabilized by pentagonal columns for operation in the North Sea. Technologically the most advanced drilling rigs at that time, they were designed to operate in water depths in excess of 1,000 feet and to drill to depths of more than 20,000 feet under the severest weather conditions. Pel–Lyn built the rigs, known as Venture One and Venture Two, for a total cost of $75 million and contracted them to major oil companies, primarily Conoco and Amoco.[32]

The successful collaboration with Dixilyn eventually led Panhandle to acquire the contract drilling firm. In 1977, Panhandle shareholders voted in favor of an exchange of stock, and the firm consequently became a wholly owned subsidiary. At

[30] Panhandle Eastern, Proxy Statement, April 7, 1977.
[31] John LaForce, interview with Clarance M. Smith, December 2, 1993, PECA.
[32] "Dixilyn: Panhandle's Latest Acquisition," *Panhandle Magazine*, no. 2, 1977, p. 1.

this time, Dixilyn was operating five offshore drilling rigs: two in the North Sea, and three in the Gulf of Mexico. The new subsidiary then purchased a sixth rig that operated in offshore West Africa.

Dixilyn further diversified Panhandle, although this was not a conglomerate investment in the style of Y&O. Both the sub and parent were energy companies. Even though the North Sea offshore drilling business was far removed from domestic gas transmission and production, there were technological, economic, and political similarities. Panhandle president O'Shields saw the acquisition as a product of the changing global energy market:

> Much of the future oil and natural gas the country and the world needs must necessarily come from offshore areas, many of which are relatively unexplored. Our own studies and evaluations by other reliable industry sources indicate that worldwide demand for offshore rigs will continue to grow for several years. This circumstance, coupled with Dixilyn's experienced management team, which will remain with the organization, along with the existence of a veteran, skilled work force, makes this an appropriate investment for Panhandle Eastern.[33]

Panhandle was successful in working with Dixilyn, and this encouraged it to expand its drilling business. Within a year, Panhandle acquired Field Drilling Company. Headquartered in San Antonio, Texas, Field was primarily an onshore drilling company, and a good business mate for Dixilyn. Panhandle subsequently merged the two drilling firms into one business unit, Dixilyn–Field Drilling Company. Eventually, Dixilyn–Field was operating more than forty land rigs and an average of 11 offshore rigs.[34]

As a Panhandle subsidiary, Dixilyn–Field contributed significant earnings to the consolidated financial performance of the company, but the 1981 downturn in oil prices brought an end to this diversification experiment. Then, the demand for contract drilling dropped almost overnight. The tremendous expense of the fifty or so drilling rigs in the Dixilyn–Field fleet was too much of a financial burden for Panhandle, which decided to sell these assets. Panhandle's experience in the con-

---

[33] Ibid.
[34] Dixilyn–Field Drilling Company, marketing catalogue, PECA.

tract drilling business was thus profitable, but short-lived. Like most entrepreneurial investments, its long-term development potential was significantly less than its short-term profit prospects.

The public/private search for gas

Both public officials and private businessmen knew that these various efforts would not solve the fundamental problem of gas shortage. For a time, it appeared that the FPC might change course in a decisive manner. The election of Richard Nixon (1969) brought to the presidency a leader whose "FPC appointments," wrote political scientist M. Elizabeth Sanders, "and legislative recommendations would support natural gas policies much more favorable to the producers than those of previous Democratic administrations."[35] But FPC chairman John Nassikas, whom Nixon appointed in 1969, was still primarily concerned about preserving the controls he thought were necessary to protect the public from unreasonable price increases. Under Nassikas's leadership, the Commission in 1970 began reexamining the area rate case hearing for the Permian Basin and Southern Louisiana to determine if higher wellhead prices might stimulate increased interstate supplies. Unable to bring itself to free the market from price controls – a decision characteristic in several regards of many of the Nixon domestic policy choices – the FPC instead allowed an increase of 30 to 50 percent on the southern Louisiana price ceiling for new gas sold under new contracts. The Commission also decided to allow producers to sell spot gas during emergency situations to the interstate pipelines at prices considerably higher than the FPC-established area rates.[36] The Commission was thus nibbling at the edges of the basic problem.

The agency continued to nibble. In 1969, it abandoned its cost-of-service rate schedule for new gas produced by transmission companies. The formula had proved inequitable, and now pipeline firms could purchase affiliated gas production at the same area rate as independent production. Panhandle Eastern's management applauded the federal policy change, recog-

---

[35] Sanders, *The Regulation of Natural Gas*, p. 125.
[36] Ibid., p. 131.

nizing that it adopted a formula akin to the fair field price Panhandle had sought in the early 1950s.[37]

The 1973 war between Egypt and Israel exacerbated the domestic gas shortage.[38] As oil supplies declined, the demand for natural gas increased. The FPC responded with a 180-day emergency sales program which allowed for unregulated prices, but the D.C. District Court struck down this rule in a case brought by a consumer coalition. Consumer advocates were – like the regulators – fixed on short-term relationships and fearful of increased prices; they were unable even to consider the possibility that a market-oriented industry would increase production, thereby increasing output and lowering prices.[39]

The FPC continued to nibble in 1975. It created a self-help program that allowed gas distribution companies to contract directly with producers. The distributors paid the producers more for temporary supplies of natural gas than federal regulators allowed them to pay their interstate transmission company suppliers. The distributor received delivery of the gas by arranging for a pipeline to transport it. Ominously, this early variant on contract carriage foreshadowed the federal transportation policy that would ultimately be adopted.[40]

---

[37] Vietor, *Energy Policy in America Since 1945*, p. 278. Also see Pipeline Production Area Rate Proceeding (Phase I), Opinion no. 326, 42 FPC 738 (1969). This opinion was affirmed in the case, *City of Chicago v. FPC*, 458 F. 2d 731 (D.C. Cir. 1971). Also see 405 U.S. 1074 (1972).

[38] Richard Vietor states that a combination of economic factors contributed to determining the timing of the shortage in *Energy Policy in America since 1945*, p. 194: "However, short-sighted public policies, consumerism, and business decision-making affected the timing, severity, and economic impact that made transition from abundance to scarcity a crisis." For an overview of the international repercussions of the rise of OPEC, see Peter Odell, *Oil and World Power*, 8th ed. (New York: Penquin Books, 1986) and John G. Clark, "The Energy Crises of 1919–1924 and 1973–1975," *Energy Systems and Policy*, vol. 4 (1980), pp. 239–71. In "Energy: A Federal Oil Firm," *Time*, February 24, 1975, p. xx, it was reported that independent marketers "suspect the major oil companies have contrived the shortage to force them out of business. . . ."

[39] Vietor, *Energy Policy in America since 1945*, pp. 289–90. *Consumer Federation of America v. FPC*, 515 F. 2d 347.

[40] Tussing and Barlow, *The Natural Gas Industry*, p. 112. The first emergency rules for transportation were issued in Order 533, 54 FPC 21 (1975).

Another means of stimulating production for interstate commerce was the advance payment program. Advance payments were essentially interest-free, five-year loans, made by interstate pipeline companies to producers to finance increased exploration; the value of the loans could become part of the pipeline's rate base. The loans were to be paid back within five years whether or not the drilling was successful, and the pipeline which loaned funds had the right to purchase any gas produced at regulated prices. In conjunction with this policy, the FPC authorized a $43.6 million, five-year program for the exploration and development of new sources of gas by Panhandle Eastern.[41]

The majority of interstate transmission companies participated in the advance payment program in 1972. Commitments in 1972 alone accounted for more than $1.6 billion. Natural Gas Pipe Line advanced more than $260 million to producers between 1971 and 1974. During the period 1970–6, Panhandle Eastern advanced about $320 million to nonaffiliated producers. The FPC allowed firms which made advance payments to include those values in their rate bases.[42]

To take full advantage of this new program, Panhandle organized Pan Eastern Exploration on January 1, 1973. The pipeline transferred its remaining gas production properties in the Anadarko basin (544 gas wells) to the new entity. Pan Eastern launched new exploration ventures and participated with a group of companies, headed by Sun Oil, in the acquisition of other new gas properties. The group successfully bid on thirteen federal offshore Gulf of Mexico tracts which the Department of the Interior had offered for lease.[43]

Although the advance payment program generated some additional production revenue and gas supply, it was short-lived. The New York Public Service Commission challenged the prac-

---

[41] Tussing and Barlow, *The Natural Gas Industry*, p. 110; Vietor, *Energy Policy in America since 1945*, p. 278.

[42] Panhandle Eastern, *Annual Report, 1977*, p. 2.

[43] Richard Vietor argued that "On the supply side, depletion pushed oil and gas exploration offshore to the continental shelf and into Alaska and other western wilderness areas." Vietor, *Energy Policy in America Since 1945*, p. 193. Also see Davis, *Energy Politics*, pp. 127–8 for discussion of the Outer Continental Shelf Act of 1953 and related issues.

tice in a lawsuit. Ratepayers objected to providing interest-free loans to producers, many of whom were major oil companies, and the federal court directed the FPC to review the merits of the program. In 1975, the FPC responded by terminating the program.

Meanwhile, the agency was trying again to set prices for natural gas that would stimulate renewed exploration. In 1974 the FPC replaced its area pricing formula with a single national price that was not cost based. The initial nationwide price for new gas was $.42 per Mcf.[44] The FPC subsequently tried to inject hints of market forces into its pricing formulas. In an attempt to mimic the market, the FPC created a three-tiered pricing formula allowing higher prices for recently produced gas: it established a price of $1.42/Mcf for gas produced during 1975–6; $1.01 for gas produced in 1973–4; and $.52 for old gas produced.[45]

Weak as these initiatives were, the new public policies and the company's manifest need for new gas encouraged Panhandle Eastern to search for additional sources of gas. It looked, for instance, to the Gulf of Mexico, where it organized a joint venture with NGPL. The venture was to take advantage of the estimated 3 Tcf of gas located beneath seventeen offshore blocks of acreage. Of the total, 1.1 trillion was dedicated to Panhandle's Trunkline and 1.3 trillion to NGPL. The United

---

[44] FPC Opinion no. 699, 51 FPC 2212 (June 21, 1974), set a single national ceiling price for natural gas.

[45] National Rates for Jurisdictional Sales of Natural Gas, Docket no. RM75-14, Opinion no. 770 (Washington: FPC, 1976); Vietor, *Energy Policy in America Since 1945*, pp. 282, 287. Also, see McCraw, *Prophets of Regulation*, p. 235; and Vietor, *Contrived Competition*, p. 114. Also see, Alfred E. Kahn, "Economic Issues in Regulating the Field Price of Natural Gas," *American Economic Review* 50 (May 1960), 507–13; and FPC Opinion 468, 34 FPC 159, p. 329. Kahn, who had earlier suggested implementation of a two-tier pricing structure, one price for old gas and one for new gas, was not at this time in favor of natural gas decontrol. In 1974, while serving on the New York State Public Service Commission, Kahn stated that gas production was "inexorably declining" and that decontrol would transfer "tens of billions of dollars" from the producers to the consumers; Sanders, *The Regulation of Natural Gas*, p. 148. Although the regulatory tide was now swinging in favor of the producers, debate intensified on the best methods to both stimulate production but prevent prices from skyrocketing.

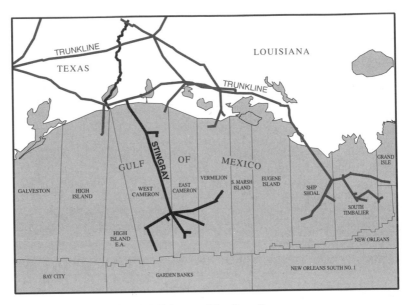

Map 8.1 Stingray Pipeline Company

Gas Pipeline Company bought the remainder and transported it through the "Stingray" system on a fee basis.[46]

Preliminary discussions between Trunkline and NGPL about Stingray had begun in 1971, just as exploratory drilling in the western Gulf of Mexico was getting under way. At that time, no hard data was available on the availability of offshore supplies. But in April 1972, the two companies had signed an agreement for the ownership of the Stingray offshore transmission system. Trunkline designed, constructed, and operated the network, most of which was built during the summer of 1974 (Map 8.1). Two of the world's largest marine construction companies, Brown & Root, Inc. and J. Ray McDermott & Company, Inc., built the system, including the offshore compressor station (1975) known as "Stingray City." Extending 100 miles from its offshore base and standing in 200 feet of water, it was further offshore than any other compressor facility. The new offshore system connected with an existing thirty-six-inch pipe-

[46] The Blue Water name already applied to a portion of the existing NGPL system; Stan Wallace offered the name Stingray.

line located twenty-seven miles out into the Gulf of Mexico built by NGPL in 1971.[47]

In 1973, the two Stingray partners brought Pennzoil, owner of United Gas Pipeline Company, into this operation, and together the three firms contributed significantly toward alleviating the shortage in the Great Lakes region. Several distribution companies, including Panhandle customers such as Consumers Power Company and Battle Creek Gas of Michigan, Citizens Gas in Indianapolis, Central Illinois Light Company, as well as NGPL's Chicago area customers benefitted from the Stingray gathering system. They commended Panhandle Eastern for its efforts to increase supply in an era of diminishing production, but of course no single offshore system could eliminate the national shortage.[48]

Alaskan natural gas

The problems of supply also sent Panhandle to Alaska, where the largest single discovery of crude oil and natural gas ever made on the North American continent had occurred (1968) at Prudhoe Bay on the North Slope. The estimates for those proved recoverable natural gas reserves ranged from 26 Tcf to

---

[47] A Federal Grand Jury later indicted Brown & Root, Inc. and J. Ray McDermott & Co., Inc., for the firm's pricing and marketing practices. This investigation resulted in an indictment of both Brown & Root and J. Ray McDermott in which the U.S. Department of Justice fined each firm $1 million. The indictment charged that the two firms "got together to decide which firm was to win specific jobs, particularly in the Gulf of Mexico, and arranged for the firm not assigned the job to submit 'intentionally high' bids." "McDermott and Brown & Root Fined," *Engineering News–Record* (December 21, 1978), p. 39. Also see, "Grand Jury Probes Brown & Root," *Engineering News–Record* (December 1, 1977), p. 17; "Offshore Contractors Sued in Civil Antitrust Action," *Engineering News–Record* (December 20, 1979), p. 46.

[48] Remarks by Robert D. Hunsucker at press briefing for Stingray system, Houston, TX, August 13, 1974, PECA. Pennzoil, owner of United Gas Pipe Line Company, had filed an application in 1973 to construct an offshore system virtually parallel to Stingray. Fearing that a competing application and regulatory complications would delay the movement of offshore gas for two years or more, NGPL and Trunkline officials entered into discussions with United Gas management to resolve the conflict. An arrangement reached during the summer of 1973 provided for United to transport up to 200 Mcf/d through the Stingray system.

Figure 8.2 Stingray City under construction

33 Tcf.[49] Arctic Gas Pipeline Company, a consortium of sixteen U.S. and Canadian firms, was the first to file a certificate application, in March 1974, to import gas from Prudhoe Bay. Its forty-eight-inch line would extend southward through the Arctic National Wildlife Range, and the $8.1 billion project quickly invoked the opposition of environmental groups. They contended that the line would endanger wildlife.[50]

Other competing proposals followed. For its part, Panhandle Eastern had already formed a feasibility study group to investigate participation in an Alaskan gas importation venture. Subsequently, Panhandle decided to enter into partnership with

[49] Tussing and Barlow, *The Natural Gas Industry*, pp. 78–86; and Vietor, *Energy Policy in America since 1945*, pp. 297–9. Also see Christopher Gibbs and Richard Vietor, "The Alaskan Natural Gas Pipeline" (Boston: Harvard Business School, Case 9-381-195, 1981).

[50] See Edward R. Leach, "Arctic Energy: In Search of a Route," *Pipeline and Gas Journal*, July 1971, p. 52; Nicholas P. Biederman, "Arctic Gas: Potential Looks Good, Many Are Studying How to Move It," *Pipeline and Gas Journal*, July 1972, pp. 21–3; Davis, *Energy Politics*, p. 136; and Walter J. Mead, *Transporting Natural Gas from the Arctic: The Alternative Systems* (Washington, D.C.: American Enterprise Institute for Public Policy Research, 1977).

five other firms to form the Northern Border Pipeline Company
(1974). The other partners were Columbia Gas Transmission,
Michigan Wisconsin Pipeline Company, Natural Gas Pipeline
Company, Texas Eastern Transmission Company, and Inter-
north, Inc. (formerly Northern Natural). This group estimated
the cost of constructing a pipeline to the U.S.–Canadian border
to receive Alaskan gas to be $1.3 billion.[51]

The environmental struggle that followed effectively killed
Panhandle's plans to use Alaskan gas to relieve the U.S. shortage. Instead, the company settled for the forty-two-inch, 823-
mile Northern Border pipeline, which extended from the Canadian border in Saskatchewan to a connection with Northern
Natural in Ventura, Iowa (Map 8.2). The line transported Canadian gas produced in Alberta. Panhandle invested a total of $86
million in the Northern Border project through 1981. Although
it proved profitable, like so many of the undertakings launched
during these years, it would not realize its full potential.[52]

Algerian LNG

Still, the need for gas was so pressing and the long-term prospects seemed so favorable, that firms like Panhandle found it
hard to resist the temptation to press on in the search for new
sources of gas. One of these involved a level of risk and liability
that would probably have frightened off an investor/CEO with
the personal perspective of William Maguire. But it did not
deter Panhandle's executives in the hectic 1970s.

Liquefied natural gas (LNG) attracted Panhandle's interest

---

[51] For a discussion of Canadian oil and gas developments and issues see John F.
Helliwell, *Oil and Gas in Canada: The Effects of Domestic Policies and World
Events* (Toronto: Canada Tax Foundation, 1989). As the plans to import Alaskan gas became public, Congress, wary of the technological problems, regulatory delays, and cost-overruns of TAPS, implemented a legislative time table
for the construction of one of the competing lines. In the process, Congress
passed the Alaska Natural Gas Transportation Act of 1976 (ANGTA, 15 USC
719) in September. Also see "Recommendation to the President Alaskan Natural Gas Transportation System," 58 FPC 810.

[52] "Panhandle Eastern Corp.," *Houston Post*, January 20, 1983. For a review of
early U.S. and Canadian relations regarding natural gas imports see Ralph
S. Spritzer, "Changing Elements in the Natural Gas Picture," in Keither C.
Brown, ed., *Regulation of the Natural Gas Producing Industry* (Baltimore:
The Johns Hopkins University, Resources for the Future, 1972), pp. 114–16.

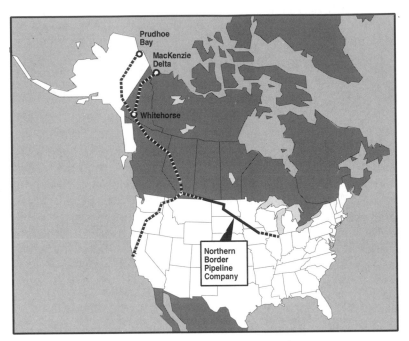

Map 8.2 Alaska natural gas transportation system

in the early part of the decade. The process of liquefying gas had
been developed by Godfrey Cabot in 1914. By lowering its temperature to −260° F, it could be reduced in volume 600 times. In
this reduced state, LNG was potentially volatile, but it could be
economically transported by special ocean tankers. LNG technology developed gradually after World War II. Despite several
catastrophic accidents, LNG technology improved gradually after World War II and this new branch of the industry continued
to grow.[53] By 1973, LNG was being transported from Brunei,
Libya, southern Alaska, and Algeria to continental Europe, the
United Kingdom, Japan, and the United States.[54]

[53] For a description of one tragic accident related to LNG importation see
Castaneda and Pratt, *From Texas to the East*, pp. 202–4.
[54] Edward Faridany, *LNG: 1974–1990,* "Marine Operations and Market Prospects for Liquefied Natural Gas" (London: Economist Intelligence Unit, Ltd.,
1974), p. 20. For a valuable analysis of the international context of energy
use see John G. Clark, *The Political Economy of World Energy: A Twentieth*

Panhandle Eastern's managers considered the possibility of importing LNG from Algeria as early as 1971 and organized Trunkline LNG Company (TLC) to conduct negotiations. By that time, Algeria was already negotiating with El Paso Natural Gas. Algeria's state-owned enterprise (SOE), Sonatrach operated the country's natural gas pipelines and its liquefaction plant in Arzew, Algeria.[55]

On December 7, 1972, Sonatrach and Panhandle signed a LNG contract. William C. Keefe, chairman of Panhandle's board, led the negotiating team. The agreement, which was subject to the completion of a firm purchase contract, provided for the delivery of 450 MMcf/d for a twenty-year period, a total of 3.3 Tcf, to commence in 1979. Sonatrach would handle all production and liquefaction operations and would supply half of the tankers needed for the project.

FPC hearings were completed in December 1976, during the "winter of the century." Concerned about frequent "brownouts" and "blackouts," the Commission quickly authorized the project.[56] During the same year, the U.S. and Algerian governments also approved the plan. The FPC agreed to allow Trunkline to sell the LNG on a favorable "rolled-in pricing" basis, using a weighted average price based on the price of both LNG and domestic gas.[57]

Panhandle had to construct a receiving terminal, provide two cryogenic tankers, and install a pipeline to connect Trunkline's main transmission system with the LNG plant. The company chose a site forty-five miles from Trunkline's main transmis-

---

Century Retrospective (Chapel Hill: The University of North Carolina Press, 1990), pp. 274–311.

[55] See Tussing and Barlow, *The Natural Gas Industry*, pp. 65–71. For background on LNG, see Vietor, *Energy Policy in America since 1945*, pp. 295–7. Originally, the Algerians estimated that it would cost $630 million to construct their plant; construction problems drove the actual cost closer to $2.3 billion.

[56] For background, see, "Is LNG the Answer?," *The American Legion Magazine*, August 1977, pp. 20–38; Les Gajay, "LNG–Import Limits and Project Subsidies are likely to be Axed in Carter Proposal," *Wall Street Journal*, April 14, 1977; James P. Sterba, "Gas Shortage a Fundamental, Long-Term Economic Threat to U.S. Experts Say," *The New York Times*, February 22, 1977; and Tim Metz, "Gas Shortages Give Some New Ammunition to LNG Advocates," *Wall Street Journal*, February 14, 1977.

[57] Opinion no. 796, 58 FPC 725; and Opinion no. 796-A, 58 FPC 2934–46.

Figure 8.3 Signing the LNG contract in Algiers, 1972 (left to right: Louis Begley, Dwight Seely, William Keefe, and Slimane Bouguerra)

sion system at Lake Charles, Louisiana, for the terminal and regasification facilities. These included a special dock for the tankers, unloading equipment, and three storage tanks each capable of holding 600,000 barrels of LNG at an estimated cost of $130 million. The double-walled, insulated tanks had to maintain LNG temperatures of −260° F. To finance the LNG tankers, Panhandle incorporated Pelmar, a wholly owned subsidiary. Pelmar, and subsidiaries of General Dynamics Corporation and Moore–McCormack Resources, Inc., in turn organized Lachmar to construct, own, and operate the two cryogenic tankers. General Dynamics and Pelmar each owned 40 percent of Lachmar, and Moore–McCormack had 20 percent. Pelmar agreed to pay $64 million for its share of the two tankers – the *Lake Charles* and the *Louisiana,* each of which cost approximately $190 million for General Dynamics to build. Construction costs were also subsidized by federal funds, long-term bonds guaranteed by the U.S. under Title XI of the Merchant Marine Act of 1936. Construction was completed in 1980.

CNAN, the Algerian national shipping company, agreed to provide three additional tankers, giving the entire fleet a total capacity of 165 Bcf per year.[58]

This was a very large, complex, international project with long-term contracts that left Panhandle vulnerable to price changes in several markets. Problems soon began to develop on a number of fronts. Costs escalated rapidly. In September 1978, Trunkline LNG estimated that it would spend $240 million on its terminal facilities. By February 1980, these costs had risen to $430 million and the final figure was $567 million.[59] Other costs were increasing as well during this era of high inflation. By 1979, management was deeply concerned that the price of LNG was going to be much higher than that of domestically produced natural gas. The cost to Panhandle derived in part from an escalator clause based on the price of No. 2 and No. 6 fuel oil at New York. As oil prices went up, Panhandle's position steadily worsened. Management explained that "the Company's commitment to LNG is based upon one paramount consideration, namely, that the gas reserves involved cannot be duplicated from domestic sources and will contribute importantly to overall supply for the balance of this century."[60] But there was cause for concern.

The management's confidence at this point appears to have been based in part on the fact that Algeria had invested about $12 billion in LNG as of 1979. This investment, the firm said, "provides a powerful counterweight to unilateral disruption."[61] But by 1980, Panhandle's confidence was wearing thin. The company publicly reported for the first time its concern over the possible adverse financial impact on the firm of an ex-

[58] Panhandle Eastern, *Annual Report, 1976*, p. 13. In February 1978, Dwight Seely was named president of Trunkline LNG Company. Seely had begun working at Trunkline in the gas supply department in 1962, He was named vice-president of gas supply in 1966 and a vice-president of Panhandle Eastern in 1976. Later, Seely would become chairman and CEO of Trunkline LNG. In February 1979, Ray Newsom, formerly chief of staff at WyCoalGas, assumed the role of Trunkline LNG president and chief operating officer.

[59] Panhandle Eastern, *Annual Report, 1979*, p. 14; and Panhandle Eastern, *Annual Report, 1981*, p. 18.

[60] Panhandle Eastern, *Annual Report, 1979*, p. 15; and Panhandle Eastern, *Annual Report, 1977*, p. 16.

[61] Panhandle Eastern, *Annual Report, 1979*, p. 15.

tended delay or project failure.[62] Sonatrach was pressing to invoke the escalator clause and increase the LNG price. Panhandle reported to its stockholders that "a Sonatrach official expressed a desire to have discussions on the LNG price and possible allocation to other customers of Sonatrach of certain volumes to which Trunkline LNG is entitled."[63] Panhandle responded that the agreement, approved by both the U.S. and Algerian governments, "does not provide for modifications of price prior to first regular delivery and has no provision for any change in quantity."[64]

There was little to encourage Panhandle in either the international or domestic news at this time. Sonatrach and El Paso Natural Gas had reached a "stalemate" in the spring of 1980 over the LNG price for their project. The Algerian SOE then suspended all LNG deliveries to that firm. The U.S. government entered the negotiations, without success, and in 1981, El Paso abandoned its LNG importation operation. The company wrote off a $365 million uninsured investment in its tanker fleet.[65] Unlike El Paso, which had not built its own receiving terminal, Panhandle's investment in its terminal could only be recovered if the plant became operational and was placed in the company's rate base. Meanwhile, the federal policies upon which the LNG venture had been based were changing as well.

A new federal energy policy

President James E. Carter, who took office in January 1977, had pledged in his campaign to solve America's energy problems. Carter proposed the creation of a Department of Energy

[62] Panhandle Eastern, *Annual Report, 1980*, p. 20.

[63] Ibid.

[64] Ibid. In 1980, Trunkline estimated that the initial price of regasified LNG placed into its pipeline system would be $6.62 per MMBtu for deliveries made before July 1, 1981. The price of Trunkline's regasified LNG was based on three elements: the LNG delivered cost (FOB) to the LNG tankers at the Algerian port, the costs of transporting the LNG across the ocean by tanker, and the capital and operating costs of operating the Lake Charles facilities.

[65] Richard H. K. Vietor, *Contrived Competition: Regulation and Deregulation in America*, (Cambridge: The Belknap Press of Harvard University Press, 1994), p. 130. Also see Robert D. Hunsucker, interview by Christopher J. Castaneda and Clarance M. Smith, PECA, who discussed Panhandle's response to the termination of the El Paso project.

(DOE), which would include all of the nation's energy-related programs, and he kept his promise to make the availability of natural gas one of his first priorities.[66] He declared the "moral equivalent of war" on energy problems. Carter supported federal efforts to revitalize the coal industry, to fund synfuel projects, and to encourage the importation of foreign produced gas. These measures were designed to improve the United States' long-term energy supply outlook.[67]

Carter and the legislature had to respond quickly to the suffering caused by the severe weather of the winter of 1976–7. The major gas producing states – primarily Texas, Oklahoma, Louisiana – consumed 41 percent of annual nationwide *intrastate* gas sales. The *interstate* gas sales were another matter. Due to restrictive FPC pricing formulas, producers had simply stopped developing interstate gas reserves. When the severe winter of 1976–7 hit, many interstate pipelines could deliver only 75 percent of their firm contract demand; other pipelines were in worse shape. In northeastern and midwestern gas consuming states, the shortage forced the shut-down of more than 4,000 manufacturing plants and hundreds of schools; more than 1 million workers were laid off.[68]

In early 1977, James Schlesinger, Carter's future Secretary of Energy, learned that Richard O'Shields, as chairman of the Interstate Natural Gas Association of America (INGAA), had created a task force led by Harry Welch (Panhandle's General Counsel), to analyze the current gas situation and identify options available to interstate pipelines in the event of such a massive shortage. Schlesinger asked O'Shields to meet with him at the White House on the day after Carter's inauguration and "present recommendations for dealing with the developing gas shortages. In the meantime, O'Shields consulted with industry officials and drafted recommendations to alleviate the shortage. At the White House meeting with Schlesinger and key members of Congress, the Panhandle Chief presented the recommendations. Within two weeks, these recommendations became the basis for the Emergency Natural Gas Act. This Act required interstate pipeline companies to make emergency de-

---

[66] Vietor, *Energy Policy in America since 1945*, p. 322.
[67] James E. Carter, *Keeping Faith: Memoirs of a President* (Toronto: Bantam Books, 1982); Davis, *Energy Politics*, pp. 136–8.
[68] See Tussing and Barlow, *The Natural Gas Industry*, pp. 113–14.

liveries to other interstate pipelines and distributors for customers in shortage areas. The act also authorized emergency purchases at prices above FPC-regulated levels.[69]

In the midst of the shortage, O'Shields appeared on the MacNeil–Lehrer News Hour in late January 1977, to discuss that situation with economist Alfred Kahn, chairman of the New York Public Service Commission, and Richard Dunham, who chaired the FPC. O'Shields said the transmission companies liked the emergency legislation, but he went on to point out that there was a more general problem stemming from the federal regulations: "There is a real natural gas shortage which has been growing for nearly eight years, and it has reached the point where the shortage we now have combined with this extremely cold weather is causing hardships in many areas. It's true that the shortage has been confined primarily to the interstate natural pipeline systems because of federal price regulations."[70] Alfred Kahn agreed: "I was one of the 'architects' of regulation, but I think it's bankrupt now. If we had deregulated, we would have had more gas; we also would have had more customers hooked up with that gas. We would not have had the restrictions that we've had in the last few years."[71]

O'Shields and other leaders in the industry urged Congress to deregulate natural gas, but Carter opted to revamp the price controls instead with the goal of stimulating production and the flow of gas through the interstate system. Dick O'Shields spoke out publicly against this policy of nibbling at the problem – but to no avail. The administration's proposals were enacted as the Natural Gas Policy Act of 1978 (NGPA), and the companion Power Plant and Industrial Fuel Use Act (1978) which discouraged natural gas use by electric utilities and large industrials.[72]

---

[69] Emergency Natural Gas Act, Public Law no. 95-2, February 1977. The delivery–transportation provision expired on April 30, 1977, while the emergency purchase authorization continued until July 31, 1977. E. O. Nelson, Panhandle's VP of Operations, traveled to Washington to assist with the administration of the Act.

[70] Transcript of "Natural Gas Supply in the United States," *The MacNeil–Lehrer Report*, Public Broadcasting Service production, January 31, 1977, p. 6, PECA.

[71] Ibid., p. 18.

[72] The resulting gas glut forced Congress to rescind the provisions of this Act which discouraged natural gas use.

The NGPA established nine major price categories, each of which contained subcategories, based on well depth, vintage, and source. Newly discovered on-shore gas received a price of $1.75 per Mcf, plus an inflation escalator which would remain effective until January 1, 1985. At that time all new gas produced for interstate and intrastate commerce would be deregulated. Old gas would be priced in three tiers from $.29 through $1.45, plus inflation, and remain regulated for an indefinite time period. The Act also created a new agency, the Federal Energy Regulatory Commission (FERC), which assumed most of the duties of the Federal Power Commission.[73]

The NGPA, which marked the beginning of natural gas deregulation in the United States, deserves most of the criticism it has received. As one prominent scholar has argued, the process of finding a regulatory solution to the shortages "could not have been more inept."[74] The new law compromised between a consumer-oriented regulatory system that reacted to short-term economic conditions with rigid price controls and a market-oriented system that worked successfully in most other industries. While the NGPA did stimulate gas production, it could not do the same for demand. A gas "bubble" – surplus deliverable gas – quickly developed which would pose an historically unique problem to this regulated industry.

The NGPA embraced a contradiction that would create major problems for natural gas pipeline companies. On the one hand, it continued the policy of encouraging companies to develop alternative sources of fuel – coal gas, Alaskan gas, LNG, even though the products would not be economically viable without government subsidies. But on the other hand, the NGPA actively sought to stimulate domestic gas production, a process that would result in lower prices and an even larger gap between the supplemental and the natural fuels. It was not at all clear that the government would then be prepared to subsidize the fuels it had encouraged companies to develop. The NGPA

[73] Natural Gas Policy Act, 1978, Public Law no. 95-621, November 1978. U.S. Congress, Senate Committee on Energy and Natural Gas Resources, *Natural Gas Pricing Proposals of President Carter's Energy Program* (Part D of S. 1469), 95th Cong., 1st sess., June 1977, pp. 235–45. *Congressional Record*, September 22, 1977, pp. S15322–65 and September 26, 1977, pp. S15595–604. Also see Vietor, *Energy Policy in America since 1945*, p. 311.

[74] Vietor, *Energy Policy in America Since 1945*, p. 312.

caught Panhandle and other pipeline companies midway in the process of supplemental fuel development, leaving several of them in a perilous position. Indeed, Panhandle took great pride in responding to the shortages by investing in coal gasification, Alaskan–Canadian gas importation ventures, and, in particular, LNG. If the new federal program was able to convert a shortage into a surplus of natural gas, the pipelines could look forward to a decade or more of major restructuring.

9

Struggle for corporate control

At 3:00 P.M. on September 25, 1982, Panhandle received its first delivery of liquefied natural gas (Map 9.1). The 900-foot Algerian tanker, the *Moruad DiDouche,* sailed past the Calcasieu jetties to a temporary anchorage in a special zone just outside the Lake Charles, Louisiana, ship channel. Early the next morning, after a routine safety inspection by the Coast Guard, the ship docked at the Lake Charles plant with a cargo of 125,000 cubic meters of liquefied natural gas. It took four days to unload – normally a twelve hour job – because extra time was needed to lower the temperature of the terminal piping and storage facilities. Trunkline anticipated another delivery from the U.S. tanker *Lake Charles* on October 5 and expected to receive shipments once every six days.

What should have been an occasion to celebrate would become in the years ahead a dark moment in the history of the firm and the industry. By the time the *Moruad DiDouche* arrived, the U.S. business system was already in serious trouble. Many American companies were losing out in competition with foreign firms. Business expenditures for research and development had declined, oil prices, at home and abroad, had increased, and interest rates had reached record levels. A new round of federal regulatory reforms began the process of un-

215

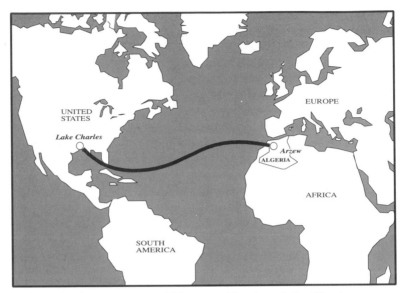

Map 9.1 Trunkline LNG route: Algeria to Lake Charles

Figure 9.1 The *Lake Charles* unloads LNG at Trunkline's Lake
Charles, LA, LNG plant

tangling business from the legacy of restrictive New Deal regulations.[1]

In 1981, the new presidential administration of Ronald Reagan promised the citizens of the United States a "New Beginning" and an end to economic malaise. Reagan's team moved to cut taxes, lower interest rates, and spark the economy. At the same time, the administration actively supported deregulation of the nation's industries with the goal of increasing the competitiveness of American business, domestically and internationally. Heightened global competition and the inefficiency of American business accelerated the structural adjustment by U.S. companies.[2]

Out of this economic turbulence, an intense merger and acquisition movement swept through the economy as corporations became part of a new market – the market for corporate control – in which a wide variety of market transactions shifted corporate control among managers, stockholders, politicians, board members, consumers, and employees. These conditions encouraged de-conglomeration. To compete more effectively against both domestic and foreign businesses, American firms began to concentrate on their core business functions and spin off diverse business activities.[3] They began as well to de-staff, seeking through decentralization to encourage innovation and achieve greater efficiency. Firms that did not change were absorbed by more efficient and powerful competitors or were driven out of business entirely. The new competition touched Panhandle Eastern, as it did most other U.S. corporations.[4]

---

[1] Louis Galambos and Joseph Pratt, *The Rise of the Corporate Commonwealth*, pp. 227, 230–1.

[2] Vietor, *Contrived Competition*, pp. 16–17; Some of the deregulation acts were Motor Carrier Act of 1980; Airline Deregulation Act of 1978; the Depository Institutions Deregulation and Monetary Control Act of 1980; and the Garn–St. Germain Depository Institutions Act of 1982.

[3] Galambos and Pratt, *The Rise of the Corporate Commonwealth*, pp. 227, 230–1.

[4] The decade of the 1980s witnessed the most dramatic episode in struggles for corporate control in the twentieth century. In *The Transformation of Corporate Control*, Fligstein argues that " . . . the central goal of managers in the past hundred years has been to make sure their firms survived. To promote survival they proposed various forms of control, both inside and outside the firm. Internally, control was oriented to ensuring that organizational resources were deployed so that top management could be confident that their

Panhandle Eastern enters the new era

Panhandle Eastern entered the 1980s with a new organizational structure, new leadership, and a new strategy. In May 1981, the stockholders approved the creation of a holding company, Panhandle Eastern Corporation (PEC), which would now control all of the pipeline's common stock, as well as that of the subsidiaries and affiliates. Panhandle Eastern's new multidivisional structure included four business units: gas transmission, LNG, coal, and other activities. Each unit had to account separately for its performance, capital, and revenue requirements. This organization sharply separated the company's regulated and nonregulated businesses and was intended to avoid adverse regulatory rulings.[5]

The new holding company had a new president. Robert D. Hunsucker became PEC's first president and chief operating officer.[6] He had been with the firm since 1952, when he started

directives were being executed. Externally, this control was oriented toward establishing stable relations between competitors to promote survival of their organizations" (p. 5). Also see Gregg Jarrell, James Brickley, and Jeffrey Netter, "The Market for Corporate Control: Empirical Evidence since 1980," *Journal of Economic Perspectives*, vol. 2, no. 1 (1988), pp. 49–68. William Lazonick argues that " . . . the belief in the efficacy of the market for corporate control is inconsistent with the history of *successful* capitalist development in the United States and abroad over the past century. The history of successful capitalist development, marked by changing international industrial leadership, shows that value-creating investment strategies increasingly require that business organizations exercise control over, rather than be controlled by, the market for corporate control." William Lazonick, "Controlling the Market for Corporate Control: The Historical Significance of Managerial Capitalism," in Frederic M. Scherer and Mark Perlman, eds., *Entrepreneurship, Technological Innovation, and Economic Growth: Studies in the Schumpeterian Tradition* (Ann Arbor: The University of Michigan Press, 1992), p. 153. Also see William Lazonick, *Business Organization and the Myth of the Market Economy* (Cambridge University Press, 1991).

[5] Panhandle Eastern, *Annual Report, 1980*, p. 4. See Certificate of Incorporation, Panhandle Eastern Corporation. Filed in Delaware on January 26, 1981. Prior to the actual merger PEC guaranteed Trunkline LNG's debt in the "Trunkline LNG Assurance Letter" on April 6, 1981.

[6] Hunsucker's tenure as chief executive officer began on January 1, 1983, and he succeeded O'Shields to the chairmanship on January 1, 1988. Previously, William C. Keefe retired as chairman of the board on January 18, 1979. He was succeeded by Richard L. O'Shields, who had been Panhandle president

as a clerk in the treasury department. He gradually worked his way through positions at National Helium and Trunkline, before becoming executive vice-president and chief operating officer (1974) and president (1978) of Panhandle Eastern Pipe Line Company. As he took office, the mood in the industry was still upbeat. Analysts were still predicting that oil prices would rise to the $100 per barrel mark and that natural gas prices would "fly-up" after the deregulation of "new gas" on January 1, 1985. The expectations were for high demand, limited supply, and rising prices. Energy firms like Panhandle Eastern which had invested heavily in such supplemental fuel projects as Algerian LNG, Canadian gas, and coal gas seemed to be in a strong position.[7]

This was true even though Panhandle Eastern's large investment in supplemental fuel projects had left it more vulnerable to market and regulatory changes than it had been in the years since the late William Maguire had consolidated his control of the pipeline. As Hunsucker later recalled, the firm had organized the holding company in part to deal with this risk. They needed to protect the firm's assets from the potential failure of the LNG project. "We had a hell of a lot of liabilities with LNG and Trunkline . . . ," he said, "and we were trying to build some corporate veils in there. . . ."[8] Shortly, Panhandle would need more than veils to protect its assets.

The pivotal year: 1982

"Right there in the fall of '82," Hunsucker said later, "it just descended on us."[9] The "it" included the combined effects of deregulation on gas supply and the national recession on demand, the introduction of high-priced LNG and Canadian gas, and the strident resistance of large distributors to the inevitable resulting price increases. The long-awaited Algerian LNG entered the Panhandle and Trunkline systems in

and CEO since 1970. Robert D. Hunsucker, executive vice-president, succeeded O'Shields as president.

[7] Joseph P. Kalt, ed., *Drawing the Line of Natural Gas Regulation* (New York: Quorom Books, 1987).

[8] Robert D. Hunsucker, interview by Christopher J. Castaneda and Clarance M. Smith, October 1, 1993, PECA.

[9] Ibid.

Figure 9.2 Richard L. O'Shields and Robert D. Hunsucker

September 1982. So too did the Canadian gas, transported through the 823-mile Northern Border Pipeline. These events should have prompted some cork popping and celebratory dinners, but by the end of the year, the mood was gloomy at Panhandle Eastern.

That same year, Panhandle and Trunkline together paid $16 million to producers for gas that could not be resold but was covered by take-or-pay contracts. These long-term contracts – signed in the 1960s and 1970s when gas prices were expected to continue to rise – required pipelines to pay for a minimum volume of gas whether it was actually received or not. The take-or-pay provisions had benefited pipeline firms in an era of shortages and rising prices, but after April 1982, natural gas prices started to fall. Deregulation was encouraging discovery, bringing fresh supplies into interstate commerce, and the newly produced gas was less expensive. The large utility customers of gas pipeline firms began to turn to the spot market and to short-term gas contracts. As these customers abandoned their long-term agreements with pipelines under the economic "force majeure" clause, the pipelines remained obligated to pay producers for gas they could no longer sell.

The FERC, and Congress, was more responsive to consumer demands for lower priced gas than it was to the predicament of the pipelines. To facilitate direct transactions between producers and consumers, the FERC authorized "blanket certificates" for the pipelines which would transport this inexpensive gas. In the early 1980s, the commission gradually instituted a new and controversial policy of allowing gas pipelines to provide contract carriage – that is, transportation services – in return for relief from take-or-pay contracts.

When large utility customers discarded their long-term contracts in favor of the spot market (short-term contracts), Panhandle Eastern had no choice but to scale back its own long-term gas supply programs during the last half of 1982. The company was now on the edge of trouble, experiencing a dramatic increase in take-or-pay liabilities. Between 1982 and 1984, management estimated that the business could have a $2.7 billion take-or-pay obligation. Sales volume was declining, and customer resistance to the prospect of increased prices stemming from the LNG project was intense. In its 1982 Annual Report, Panhandle acknowledged that "the interaction of . . . weaker demand, increased rates of domestic production and the start of deliveries from two imported gas supply projects, Canada and Algeria – has created a current supply surplus which could have a serious financial and operating impact in the years immediately ahead."[10] Saddled as it was with capital-intensive supplemental fuel projects, Panhandle had substantial reason to be concerned about the company's future.

One of the primary concerns had been the repeated delays in getting the LNG project underway. Sonatrach notified Panhandle that it "had encountered delays in the completion and repair of certain facilities in Algeria. . . ."[11] A new forty-inch transmission line extending to Arzew required repairs, preventing deliveries before late October or November. Trunkline sent a team led by Wayne Perry to Arzew to inspect the damage, and they confirmed that the line was damaged and contained significant amounts of rust. Although Trunkline engineers concluded that with minor revisions Sonatrach could begin LNG deliveries in December, no deliveries were made in 1981. Panhandle became suspicious that the delays were linked to Sonatrach's demands for "upward revision of the price."[12]

In early 1982, Sonatrach officials said the LNG facilities would not be operational until April 1982 – and reiterated their interest in renegotiating the LNG price. Panhandle's representatives repeated their position that the sales contract did not provide for a price renegotiation prior to the "attainment of full regular deliveries."[13] By this time, Panhandle was in a very tight bind. Trunkline could not add its new LNG assets into the rate base to begin the process of cost recovery until after it received its first shipment. Although company officials considered breaking out of the entire deal, they allowed the venture to move forward. In the words of Del Campbell, vice-president of TLC: "Once [we] started building the plant, then

---

[10] Panhandle Eastern, *Annual Report, 1982*, p. 7.

[11] Panhandle Eastern, *Annual Report, 1981*, p. 18.

[12] Wayne Perry, interview on October 29, 1993 by Christopher J. Castaneda and Clarance M. Smith, PECA. News Release, Panhandle Eastern Corporation, August 21, 1981; October 19, 1981; and February 11, 1982. Also see Panhandle Eastern, *Minutes*, December 3, 1981, p. 6.

[13] Panhandle Eastern, *Annual Report, 1981*, p. 20.

[we were] sort of pregnant."[14] The investment was worthless unless it could be placed in the rate base.[15]

As the firm's LNG deal became increasingly problematic, the company again attracted national media attention. As *Business Week* reported in an article titled "Panhandle Eastern: A gas-shortage gamble is blowing up in its face," the pipeline firm had – like most other gas firms – sought alternative sources of gas for customers in the 1970s. Now, "the whole range of company projects that once seemed admirable – liquefied natural gas imports, synthetic fuels, and sponsorship of the Alaskan gas pipeline – is beginning to sour." The magazine was deeply suspicious of the Algerian claim that technical problems were causing the delay in shipments. "Algerians want to sell their gas at a price equivalent to oil," the magazine stated, " . . . about $5.25 per Mcf" compared to the $3.94 under the current contract. Ominously, *Business Week* reminded its readers that in 1981, El Paso Natural Gas had eliminated its operational Algerian LNG program and wrote off $365.4 million.[16]

Panhandle discovered – as had other firms over the century – that it can be financially risky to do business with nation states and their agents. The normal rules of commerce seldom apply. In July 1982, Trunkline tried to force Sonatrach to begin deliveries by filing a petition for arbitration with the International Chamber of Commerce (ICC) in Paris. In the filing, Trunkline and Panhandle Eastern said they could discover "no valid reason under the contract for the failure of Sonatrach to initiate LNG shipments." Sonatrach had begun delivering substantial volumes of LNG to Gaz de France in February 1982 "under a price agreement signed that month to a 1976 contract."[17] In July 1982, Panhandle's board for the first time officially discussed the possibility that the entire project would fail.[18]

The arbitration request seemed to spur Sonatrach into action. After receiving a shipping schedule in early August, Trunkline withdrew the arbitration petition. Sonatrach contin-

---

[14] J. Del Campbell, interview by Clarance Smith, September 10, 1991, PECA.

[15] Panhandle Eastern, *Annual Report, 1981*, p. 21.

[16] "Panhandle Eastern: A Gas-Shortage Gamble Is Blowing Up In Its Face," *Business Week*, May 24, 1982, p. 106.

[17] Panhandle Eastern Corporation, News Release, July 1, 1982.

[18] Panhandle Eastern, *Minutes*, July 28, 1982, pp. 3, 6.

ued to press for a price increase, but at least the Algerian firm now seemed prepared to deliver substantial supplies of LNG.[19]

By mid-August 1982, two of Panhandle's largest customers, MichCon and Consumers Power, were protesting that the Algerian gas was neither needed nor marketable. They suggested as well that for political reasons Algeria might not prove to be a reliable supplier.[20] Robert W. Stewart, MichCon's president, announced that his firm would take "every legal action possible" to avoid receiving high-priced gas. "We do not want the Algerian supplies," he stated, "we cannot sell it, and we will oppose taking the gas during the FERC hearings on this matter."[21] The LNG deal, he said, was "typical of the insensitivity which interstate pipeline suppliers have shown for natural gas consumers."[22] In September, U.S. Representative John Dingell (D–MI), chairman of the House Energy and Commerce Committee, backed up these charges by calling for a FERC reevaluation of the contract. Other utility companies mounted similar protests.[23]

The government steps forward

In this situation, the FERC had to act. Panhandle continued to maintain that it needed the gas to meet its obligations, but the

[19] Panhandle Eastern, *Annual Report, 1981*, p. 21; "Trunkline LNG Gets Loading Schedule from Algerian Firm," *Houston Post*, August 10, 1982; and "Trunkline to Receive Its First Algerian LNG," *Houston Chronicle*, August 10, 1982. In the original 1975 contract, the base price for the LNG was $1.30 per MMBtu delivered at Arzew to the LNG tankers. With the adjustment, the price was subject to the higher of two indices: monetary exchange rates and/or changes in the prices of No. 2 and No. 6 fuel oil delivered at New York harbor. The LNG price could vary by $.10 for every $1.00 per barrel change in the price of fuel oil. This formula resulted in an F.O.B. Algeria price of $3.94 per MMBtu as of January 1, 1982. The new agreement called for a $.17 per MMBtu change in the price of LNG per $1.00 per barrel change in the price of fuel oil.

[20] Panhandle Eastern, *Minutes*, December 1, 1982.

[21] "Customers Balk at High Price LNG," *Energy News*, August 16, 1982.

[22] "Midwest Utilities to Trunkline Gas: Stick It Up Your Pipeline!" *The Energy Daily,* vol. 10, no. 150, p. 1. Also see, Michigan Consolidated Gas Company, *News Release*, August 10, 1982.

[23] See, "Panhandle Eastern Ordered to Justify Algerian Purchase," *The Houston Post*, September 25, 1982; "Trunkline Makes Its Case for Algerian LNG," *Houston Chronicle*, September 28, 1982; and "Trunkline to Receive Its First Algerian LNG," *Houston Chronicle*, August 10, 1982.

LNG was very expensive.[24] Consumers would obviously suffer the effects of more expensive gas supplies, and their political representatives would continue to pound on the agency. The Commission was independent of, but certainly not oblivious to, political pressure. The LNG would raise overall prices in the areas served by Panhandle and Trunkline. On the other hand, the FERC could not ignore the fact that the federal government – including the Commission – had formally sanctioned the contract when it appeared that LNG would serve the public interest. With federal support, Panhandle had invested approximately $640 million in the project, and oil and gas industry analyst John Jay Jones, Jr. estimated that if Panhandle's LNG license were revoked, the company would have a pretax loss of about $510 million for 1982.[25]

In these circumstances, the Commission did just what one would expect: it waffled. When the large utility companies inundated the agency with requests to cancel the permit, FERC temporarily denied Trunkline permission to pass on its LNG costs to customers. MichCon, Laclede Gas Company of St. Louis, and other Michigan industries formed the "Association of Businesses Advocating Tariff Equity" (ABATE) and filed a request with the Economic Regulatory Administration (ERA) for the revocation or suspension of the LNG deal. FERC and ERA then agreed to conduct an "unprecedented" joint review, beginning October 28, 1982, of the entire contract. In that same month, Columbia Gas, one of Panhandle's biggest customers, cited "economic force majeure" and refused to pay the $7.5 million minimum bill for August. Columbia was only one of many distributors in the same predicament as the investigations got underway.[26] Panhandle's management downplayed the importance of federal involvement and stated that the review was no more than an effort "to clarify the status" of the

[24] "Agencies put hold on gas price boost in Panhandle deal," *Houston Chronicle*, October 29, 1982. For a brief review of international LNG operations see Ferdinand Banks, *The Political Economy of Natural Gas* (New York: Croom Helm, 1987), p. 48.

[25] "A squabble over LNG: Price Is the Issue," *Houston Chronicle*, October 1, 1982. "Panhandle Faces Balky Customers on Algerian Gas," *Wall Street Journal*, September 30, 1982.

[26] Panhandle Eastern, *Minutes*, April 27, 1982. Columbia later agreed to pay PEC's take-or-pay and renegotiated a new minimum bill.

project, but in fact the company's future was in government hands.[27]

In mid-December 1982, the Commission recommended that Trunkline go ahead with the project, largely on the basis of the sound technical and legal aspects of the original contract. FERC acknowledged that the LNG was costly but said that its primary concern was the reliability of supply. James E. Rogers, Jr., the FERC's Deputy General Counsel for litigation and enforcement, presented the agency's staff report to Curtis L. Wagner, Jr., Administrative Law Judge. Rogers said: "while the evidence is clear that we may not need the Algerian LNG in the short run – that is, over the next several years – the record shows there continues to exist a long-term need for the supplemental gas supply so long as it can be relied upon."[28] If he were to rule on a new project, Rogers said, he "would recommend rejection. . . . , in part because there is an unacceptable risk that the Algerians would interrupt LNG supplies in the future in order to gain pricing or other concessions."[29] But now Panhandle had the government on its side – at least for the moment.[30]

On Friday, January 28, 1983, Judge Wagner ruled that Trunkline could charge its customers for the LNG and set March 1 as the date the new rates would go into effect. The price was too high, he said, but Panhandle's agreement with Sonatrach was valid. The FERC accepted Judge Wagner's decision, but the issue was clearly not settled. The agency gave Panhandle seven days to explain how its LNG contract met the "public interest." Two days later, the ERA refused a request to cancel the contract, but it required Trunkline to provide a contingency plan for continuing service in the event LNG imports ceased.[31]

---

[27] "Issue of U.S. imports of Algerian LNG Heats Up," *Oil and Gas Journal*, November 8, 1982, p. 116.

[28] "Supply Reliability Seen Key Issue In LNG Case," *Oil and Gas Journal*, December 20, 1982, p. 21.

[29] Ibid.

[30] "Short Takes," *Houston Post*, December 18, 1982.

[31] "Judge's ruling allows firm here to import Algerian gas," *Houston Post*, January 29, 1983. "Panhandle Is Prodded To Pay Less for Gas But License Is Upheld," *Wall Street Journal*, February 24, 1983; and "Panhandle Eastern Wins Round in Right to Import Expensive LNG from Algeria," *Wall Street Journal*, January 31, 1983. "Algerian Gas Ruling Could Set Precedents," *Houston Chronicle*, March 3, 1983. "Ruling on Algerian Gas," *The New York Times*, February 24, 1983.

The political storm over gas prices became ever more intense. A wide array of groups including politicians who represented midwestern consumers, a number of consumer groups, and gas customers attacked the contract. Senator Thomas Eagleton (D–Mo) objected vociferously: "The benefit will go to the Algerians who will make off like bandits and to the pipelines who lined up this ridiculous deal."[32] The agency responded by once again giving Trunkline fifteen days to show cause "why its authorization to build and operate liquefied natural gas facilities at Lake Charles, La., and import LNG from Algeria should not be revoked." At this time, the LNG was entering Trunkline's system at $7.38 per Mcf up from $7.16, but the FERC was still not prepared to overturn the contract. It again permitted Trunkline to pass the higher costs through to its customers, but warned Panhandle that if it did not renegotiate the LNG contract by March 31, FERC would consider voiding the entire trade.[33]

During 1983, the FERC gradually intruded more and more into the relationship between Trunkline and Sonatrach. On the tenth day of each month, Trunkline submitted a report on the complex negotiations and the deliveries received from Algeria.[34] Panhandle officials met seven times with Sonatrach representatives in January alone in an attempt to reduce both the price and the contracted volume of LNG. In February, the Commission and the ERA both renewed their support for the contract, but Panhandle had to compromise this time.[35] The firm agreed to establish an Energy Assistance Program (EAP) to

[32] "Pact for Costly Algerian Gas Upheld," *The Victoria Advocate*, February 24, 1983.

[33] "Algerian Gas Contract under Fire," *Houston Business Journal*, July 25, 1983.

[34] "FERC Asks PL to Report on Algeria/LNG Pricing Talks," *Platt's Oilgram*, June 6, 1983.

[35] See ERA Order no. 50 of February 25, 1983 and FERC's Decision and Order of February 28, 1983. Both the FERC and the ERA required Trunkline to seek through renegotiation a reduced base price of LNG. On July 18, 1983, FERC ordered Trunkline to show cause that it had not violated its certificate. Trunkline LNG and Trunkline Gas responded on August 2 that they had not violated the certificate. On September 23, the ERA reopened proceedings related to Order no. 50. Trunkline LNG responded on October 31, 1983 by stating that the ERA did not have authority to rule on decisions made by FERC.

help low-income residential customers using Panhandle's gas. The initial program was funded up to $1.3 million.[36]

As Panhandle's problems mounted, a new threat emerged. During the first quarter of 1983, its stock price briefly fell below $25 per share, less than the break-up value of the corporation. Another entity could conceivably purchase the outstanding common stock, sell all of the firm's assets, pay all of its debts, and still realize a substantial profit. The active market for corporate control that had developed in the United States during these years made this outcome a real possibility, and management quickly decided to introduce a series of possible legal and financial responses to a possible takeover bid. In 1983, Panhandle Eastern hired Kidder, Peabody, Inc., to develop an "Immediate Action Plan" which included among other things a calculation of the current market value of the company.[37] In late January 1983, representatives of Kidder, Peabody presented "to the Board certain information and action step recommendations which might be of benefit in the event of an unnegotiated tender offer."[38] This meeting resulted in a decision by the board to amend Panhandle Eastern's certificate of incorporation to give shareholders greater assurance of fair treatment and prevent a rapid change in management without stockholder approval.[39]

---

[36] Panhandle Eastern, *Minutes*, January 26, 1983, p. 12. For an overview of energy assistance programs see Christopher J. Castaneda, *Federal Energy Assistance: An Old Deal in a New Package,*" M.A. thesis, University of Houston, 1986.

[37] See Tom Irwin, *Panhandle Eastern's Summer of 1986*, unpublished manuscript, January 1993 (Draft date 3/15/93), p. 5 (hereinafter referred to as *Summer of 1986*).

[38] Panhandle Eastern, *Minutes*, January 26, 1981, p. 1.

[39] Panhandle Eastern, *Minutes*, January 26, 1983; February 23, 1983; and April 27, 1983. As Panhandle Eastern Corporation prepared for a possible takeover, Trunkline avoided a potentially serious accident during January 1983. A problem developed in one of the Algerian LNG tankers, the *Bachir Chihani*, on its way to Lake Charles with a full load of LNG. The day after leaving Arzew, the ship, which had been in dry dock more than one year for modification of its cargo containment system, notified the U.S. Coast Guard that a leak had been detected. The ship was of the "membrane" variety, unlike the more modern spherical type built by Lachmar. The membrane ship was double hulled. Within the inner hull, a very thin membrane, sup-

Even these measures could not protect the firm and its management if something could not be done about LNG. On March 1, 1983, when Trunkline began including the LNG costs in its rates, volume quickly declined. The company's sales volumes sagged from 526 Bcf in 1982 to 300 Bcf that year. Management projected sales of only 250 Bcf in 1984. Its delivered gas costs had jumped from $2.96 to $4.69 per MMBtu, making Trunkline the highest-priced pipeline among its fifteen largest competitors in the United States. The Algerian LNG comprised 26 percent of all gas supplied to Trunkline customers and 53 percent of the company's purchased gas costs.[40] Sonatrach reluctantly agreed to reduce deliveries in April, and Trunkline put its two tankers into lay up in Newport News, Virginia.[41] In June, Trunkline received only two of its five scheduled deliveries, but each one consisted of 2.7 billion cubic feet of regasified fuel.

Panhandle continued to press Sonatrach for relief, as did other LNG importers. The two parties held numerous meetings, and the Algerian's SOE at first cut its shipments by 40 percent between April 1, 1983 and November 30, 1984.[42] As Sonatrach knew, however, the entire international LNG business was sick and it had little incentive to make further conces-

ported by plywood, insulation, and the inner hull itself contained the LNG while the outer hull served as a safety feature and could hold the LNG if necessary. As part of the safety precautions, nitrogen gas was injected between the inner and outer hulls. Methane mixed with a suitable quantity of nitrogen does not burn, but methane is detectable in the nitrogen purge. A number of "membrane" ships had problems with methane showing up in the nitrogen purge just as with the *Bachir Chihani*. The leak remained very small and resulted in no significant loss of cargo. P. A. Turek to *Lake Charles* employees, January 10, 1983, PECA.

[40] Sam Fletcher, "Expensive Gas Imports Stop," *The Houston Post*, December 15, 1983; "Panhandle Eastern Halts Its Purchases of Algerians' Gas," *Houston Chronicle*, December 15, 1983.

[41] "Algerian Gas Negotiations Still Continuing," *Houston Chronicle*, June 11, 1983; and Panhandle Eastern, *Annual Report, 1983*, p. 12.

[42] "Panhandle to Lower LNG Purchases 40 percent under Algerian Pact," *Wall Street Journal*, July 13, 1983; "FERC Finds Trunkline Deficient in LNG Talks; Warns of Import Cutoff," *Platt's Oilgram*, July 22, 1983; "Trunkline Justifies Need for Continued Sonatrach LNG Link," *Platt's Oilgram*, August 4, 1983; and "Algerian Officials Talk LNG in Washington," *Platt's Oilgram News*, July 29, 1983.

sions.[43] In October, Sonatrach informed Panhandle that it would provide no further volume or price reductions, nor would it participate in any FERC or ERA hearings.[44]

By the fall of 1983, powerful industrial and political forces were mustered against the company. Diamond Shamrock and Sun Exploration and Production were particularly adamant in their calls for a new policy. Trunkline, these producers said, was purchasing high-priced Algerian LNG instead of abundant offshore gas priced between $2.39 and $2.73 per Mcf.[45] The logic of their contentions was unassailable. In the Senate, Foreign Relations Committee Chairman Charles Percy (R–IL) introduced a bill requiring the Department of Energy (DOE) to suspend Trunkline's authorization to import Algerian LNG within ten days after enactment.[46] On November 21, President Reagan signed a less forceful but threatening bill requiring Secretary of State George Shultz and Energy Secretary Hodel to report to Congress within thirty days on efforts to "promote lower prices and fair market conditions" for U.S. imports of Algerian LNG.[47]

---

[43] "Spanish Refusing to Abide by Take-or-Pay for Algerian Gas," *Platt's Oilgram*, May 25, 1983; "Algeria's Nabi Renegotiating Gas Supply Pact with Spain," *Platt's Oilgram*, April 8, 1983; "French Gov't Unwilling to Make Up For Algerian Gas Surcharge in 1984," *Platt's Oilgram*, September 21, 1983; and "Participants at World LNG Parley Lament Market's State," *Platt's Oilgram*, May 17, 1983.

[44] Panhandle Eastern, *Minutes*, October 26, 1983.

[45] "Producers Blast Trunkline Gas Pact with Algeria in ERA Filings," *Platt's Oilgram*, October 31, 1983.

[46] "Algerian Imports," *Platt's Oilgram*, September 28, 1983.

[47] "Washington Roundup," *Platt's Oilgram*, December 1, 1983. Congress, November 1983, passed legislation providing that "the United States Government should move immediately to promote lower prices and fair market conditions for imported natural gas." Panhandle Eastern, *Annual Report, 1984*, p. 46. Panhandle's situation was presented no more clearly than when board member Donald E. Petersen, president of Ford Motor Company, resigned in part because of a conflict of interest which he described as follows: "At the time I joined the Panhandle Eastern board, I realized that care would have to be taken to avoid any conflicts resulting from Ford's status as a purchaser of natural gas from utilities and Panhandle Eastern's status as a seller to these same utilities. . . . there appears to be a fundamental incompatibility between Ford's desire to shift from natural gas to alternative fuels, and its efforts to encourage its utility suppliers to do likewise, and Panhan-

Panhandle executives began to consider options ranging from suspending the LNG imports to filing for bankruptcy. They met secretly several times to ponder the worst case scenario. "We did some significant research study," recalled Hunsucker, "if this and this and this and that occurs, when do you go to Chapter 11?. . . . We were talking about a significant corporation going under."[48] In this scenario, a bankruptcy filing by Trunkline LNG Company would drag Panhandle Eastern Corporation into bankruptcy, as well.[49]

Playing hardball overseas and at home

Trunkline LNG indefinitely suspended LNG purchases as of December 12, 1983.[50] Immediate applause came from the firm's customers (Table 9.1). "We are very pleased with this decision . . . ," said Consumers Power Company, "and believe it is the result of pressures exerted by Consumers Power and others who have maintained that the cost of LNG is too high and its supply is not presently needed."[51] Other large customers –

dle Eastern's goal of forestalling any additional loss of natural gas sales to industrial users and, indeed, to recapture if not increase such sales. This appearance of a conflict has been magnified by Ford's membership in utility customer groups which are aggressively opposing proposed natural gas rate increases, including those sought by Panhandle Eastern, as well as opposing Panhandle Eastern's LNG project." Donald E. Petersen to Panhandle Eastern board, November 3, 1983.

48 Robert D. Hunsucker, interview by Christopher J. Castaneda and Clarance M. Smith, September, 1993, PECA.

49 Harry S. Welch, interview by Clarance M. Smith, April 9, 1992, PECA.

50 During a relatively short period of operation from September 1982, to December 1983, Panhandle received 47 LNG tanker loads amounting to 120 Bcf of natural gas.

51 R. D. Hunsucker and R. L. O'Shields to Stockholders, December 14, 1983, Folder: "LNG – Lake Charles Project" Box: LNG, PECA. Also see Consumers Power Company, News Release, December 14, 1983; Folder: "LNG – Lake Charles Project" Box LNG; Also see "Panhandle Suspends 20-Year Contract To Buy Algerian LNG, Citing High Prices," *The Wall Street Journal*, December 15, 1983. During July 1984, management learned that Consumers Power Company, Panhandle Eastern's largest customer and one of the opponents of high-priced LNG, had canceled construction of its Midland nuclear project, which had been a huge cash drain. Consumers notified Trunkline that for financial reasons it could only pay for gas through June. Concern over the possibility of losing Consumers Power forever as a customer prompted Panhan-

Table 9.1. *LNG deliveries from Sonatrach^a, 1982–1983*

Month	Deliveries	Volume (Bcf)*
1982		
September	1	2.67
October	1	2.67
November	3	8.01
December	7	18.69
1983		
January	5	13.35
February	5	13.35
March	4	10.68
April	3	8.01
May	4	10.35
June	2	5.34
July	2	5.34
August	2	5.56
September	1	2.67
October	3	8.01
November–December	4	10.35

^aapproximate

Michigan Gas Utilities, Illinois Power, Central Illinois Public Service, and Northern Indiana Public Service – had apparently threatened to cut purchases to minimum levels (if not completely) in 1984. As Hunsucker and Chairman O'Shields explained: "The financial health of our transmission business and our aim to maintain the long-term viability of the LNG project made this extraordinary business decision unavoidable."[52] Af-

dle Eastern to extend it a $240 million line of credit for gas taken during the summer of 1984 since Consumers's banks were unwilling to make more loans to the struggling utility. This credit allowed Consumers Power to fill its gas storage facilities prior to the 1984–5 winter heating season and continue gas sales without interruption. Panhandle Eastern subsequently participated in a plan to convert the Midland facility into a cogeneration plant. Panhandle Eastern, *Minutes*, May 23, 1984; September 26, 1984.

[52] See, "Trunkline suspends LNG purchases from Sonatrach," *Platt's Oilgram*, December 15, 1983; and "Panhandle Suspends 20-Year Contract to Buy Algerian LNG, Citing High Prices," *Wall Street Journal*, December 15, 1983. Along with the LNG suspension, Panhandle announced that it would request

ter Trunkline placed its LNG plant in a standby mode, the only U.S. LNG import facility in operation was Distrigas Corporation's LNG terminal in Everett, Massachusetts.[53]

Sonatrach fought back. That enterprise had made large investments in both the Hassi-R'Mel gas field and three tankers to support the Trunkline project. The suspended contracts would cost Sonatrach an estimated $500 million in 1984. The Algerian SOE's management noted that during a recent visit to Algeria by U.S. Energy Secretary Hodel, they had been told that the U.S. government wanted the contracts adjusted, but not abrogated. The Algerians said that they would "defend their interests with greater vigor."[54]

Panhandle Eastern's management now faced huge liabilities and gave the bad news to its stockholders:

> ... the take-or-pay exposure could amount to $1,500,000,000 with respect to 1983, and $900,000,000 with respect to 1984, based on actual sales volumes in 1983 and projected sales volumes for 1984, and assuming no purchases of LNG from Algeria during 1984.[55]

Management gloomily added that there was "substantial uncertainty concerning the continued viability" of both pipelines.[56] In addition to the struggle with Sonatrach, the firm was burdened with a ship-or-pay agreement with Lachmar, which the latter company claimed had inflicted damages of more than $860 million.[57]

FERC approval to reduce its rates by 19 percent, or $.88 per Btu, effective January 1, 1984. "Panhandle Eastern Suspends Algerian Gas Purchases," *Dow Jones News Service*, December 14, 1983.

[53] FERC allowed Panhandle Eastern to recover Trunkline LNG's cost of maintaining the LNG facilities and its debt service in its rates through the minimum bill tariff. D. J. Campbell to Kenneth F. Plumb (FERC), February 13, 1985. Wayne Perry, interview by Christopher J. Castaneda and Clarance M. Smith, October 30, 1993, PECA.

[54] "Algerian's Vow to Defend Their Gas Interests More after Panhandle Move," *Platt's Oilgram News*, December 16, 1983; and "Algerian LNG: Sonatrach Sends Trunkline LNG Co. A Token of Its Displeasure," *The Energy Daily*, January 26, 1984.

[55] Panhandle Eastern, *Annual Report, 1983*, p. 49.

[56] Ibid., p. 41.

[57] The evidentiary hearing concluded in October. In early January, the three-member arbitration panel issued a stay order until the conclusion of Sona-

By early 1984, Sonatrach had dug in its heels. The Algerian enterprise complained about Panhandle's "bad attitude" and called on the U.S. managers "to respect your contractual obligations. . . . "[58] Sonatrach formally requested arbitration through the International Chamber of Commerce. As these conflicts intensified, the securities rating agencies lowered their evaluations of both Panhandle and Trunkline Gas. By the fall of 1984, Moodys had dropped Panhandle to BAA and Trunkline to BAA3. Standard and Poor's dropped them to BBB and to BB−, respectively.

PEC's executives looked for ways to protect the company's assets and ward off takeover attempts. One way to push the value of the stock above the potential break-up value was to restructure. They focused first on Anadarko Production Company, Panhandle's very successful oil and gas production subsidiary. They started by highlighting Anadarko in corporate publications. In the parlance of the 1980s takeover lingo, they established Anadarko as the "crown jewel" of their otherwise problematic kingdom.[59]

Panhandle's management became increasingly nervous about the possibility that their firm would become a takeover target. They had hired Georgeson & Company, Inc., an investor relations firm, to identify and analyze the types of stockholders holding Panhandle Eastern stock. In addition, the consultants provided a "stockwatch" service to monitor trading in Panhan-

---

trach's proceedings against TLC at an international arbitration in Geneva on the condition that TLC pay Lachmar's debt service obligations and other costs and expenses of Lachmar due after February 15, 1985. Sonatrach also charged both TLC and PEPL with breach of contract and demanded payment of $171 million for LNG not taken during 1983. TLC and PEPL rejected Sonatrach's claim except for $1.89 million – which it paid. Panhandle Eastern, *Annual Report, 1984*, p. 46. During early 1985, Panhandle Eastern agreed to pay approximately $32 million to Lachmar, pending the outcome of the arbitration, for the cost of laying up the two tankers and servicing the construction bond debt. "Panhandle Eastern Agrees to Pay Lachmar costs," *Oil and Gas Journal*, January 14, 1985, p. 47.

[58] D. J. Campbell and R. D. Hunsucker to Sonatrach, March 9, 1984; Assistant General Director of Sonatrach to D. H. Seely, Jr., March 22, 1984, Folder: "LNG – Lake Charles Project," Box LNG, PECA.

[59] "Panhandle Eastern Mulls Spinoff of Anadarko Affiliate," *Oil and Gas Journal*, August 4, 1986, p. 23. Anadarko Production Company, *Annual Report, 1984*, pp. 5–6.

dle's common stock, looking for early warning signs of a possible takeover attempt. Shortly, their service would prove valuable.[60]

Deregulation accelerates

While watching the stock market for signs of a takeover attempt, Panhandle's management continued to monitor carefully the policy shifts of the FERC. After encouraging the development of a "spot market" for gas, the Commission had begun in the early 1980s to issue "blanket certificates" to pipelines for transporting gas to distributors or other industrial firms.[61] Blanket certificates were followed by a new program which encouraged pipelines to form Special Marketing Programs (SMPs) under which they would buy, sell, and transport "spot market" gas for customers. These SMP affiliates could transport gas for customers who were seeking lower priced spot gas outside of their higher priced, long-term contracts with the pipelines. Panhandle Eastern formed two SMPs, PanMark and MAT. Both programs operated under temporary certificates expiring in 1985.[62]

Panhandle Eastern had also adopted an aggressive gas transportation policy to accommodate customers determined to break their gas purchase contracts with pipelines. The Central Illinois Light Company (CILCO) was angry that Panhandle, its sole supplier of natural gas, initially refused to allow it to buy spot market gas. The dispute escalated when in early 1984 the Illinois attorney general filed an antitrust suit against Panhandle. Philip O'Connor, chairman of the Illinois Commerce Commission, subsequently proposed to Congress his solution to the "virtual havoc in the natural gas industry: . . . immediately change natural gas interstate pipeline status to common carriage."[63] Although Illinois eventually lost the suit, it had

---

[60] Irwin, *Summer of 1986*, p. 5.

[61] Order 319, 3 FERC 30,477 (1983).

[62] A court decision subsequently disbanded the Special Marketing Programs. Panhandle Eastern, *Annual Report, 1983*, p. 9; and Panhandle Eastern, *Annual Report, 1984*, pp. 8–9. *Maryland People's Counsel v. FERC*, 761 F.2d 768 (D.C. Cir. 1985)

[63] *State of Illinois v. Panhandle Eastern Pipe Line Company, Central District of Illinois*, "Motion of Defendant Panhandle Eastern Pipe Line Company for Award of Attorney's Fees and Imposition of Sanctions," p. 2; *State of Illinois*

pushed Panhandle to begin "the long, hard process," recalled John A. Seiger, then general counsel, "of revising its world view and adjusting to the emerging new world order."[64]

The FERC was also moving toward the creation of a national gas market in which pipelines could offer competitive marketing, sales, and transportation services while consumers would contract directly with producers for gas supply. But the old order of restrictive gas sales contracts using minimum bills, escalator clauses, and take-or-pay agreements – gave way slowly.[65] This was true even though consumer groups, gas distributors, and a number of powerful politicians were exerting intense pressure on Congress and the FERC to make less expensive natural gas available. Responding to their demands, the Commission eliminated by way of Order 380 those contracts – so-called minimum bills – which required purchasers to maintain minimum purchase volumes, or else pay for gas not taken.[66] When this restriction was dropped, distributors began abandoning higher cost pipeline supplies and buying more and more gas on the growing spot market; by early 1985, such direct purchases from producers accounted for approximately 14 percent of the nation's gas sales.[67] Pipeline firms were now left with their gas purchase contracts with producers in full force, while consumers were free to deal directly with producers, bypassing the pipelines. Something had to give on the large take-or-pay contracts the pipelines were still required to honor.[68]

The FERC attempted to remedy this situation, but it moved far too slowly. Order 436 sought to make transportation services a distinct business function. It provided for nondiscriminatory transport service by the pipelines; promised reviews of

   v. Panhandle Eastern Pipe Line Company, Central District of Illinois, "Memorandum in Support of Defendant Panhandle Eastern Pipe Line Company for Award of Attorney's Fees and Imposition of Sanctions," p. 3.

[64] John A. Seiger to Christopher J. Castaneda, January 30, 1995.

[65] Vietor, *Contrived Competition,* p. 144. Also see David J. Teece, "Vertical Integration and Dual Distribution in the Natural Gas Industry: Causes and Consequences," December 29, 1986.

[66] FERC, Docket RM83-71-000, Order 380, Final Rule, May 25, 1984, pp. 11–14.

[67] Vietor, *Contrived Competition,* p. 146.

[68] For background on direct sales see Robert C. Means and Robert S. Angyal, "The Regulation and Future Role of Direct Producer Sales," *Energy Law Journal* vol. 5, no. 1 (1984), pp. 1–46.

take-or-pay buyouts for the pipelines that adhered to 436; and imposed additional rules on the pipelines to prevent them from shifting resources between their competitive and noncompetitive services. Utility customers could now reduce their existing pipeline contract demand to zero over a five-year period.[69]

Order 436 left Panhandle's management upset because it did not force a renegotiation of the existing gas purchase contracts. Hunsucker later commented: "It takes two to tango. Our pipeline – any pipeline – cannot lower its contracted gas price without agreement from the producers involved. Those producers will not dedicate long-term reserves to us at a price equivalent to spot gas, which is generally sold short-term on an interruptible basis. We, therefore, cannot purchase or sell at spot prices over the long term. So much for economic theory." The FERC's new policies placed additional pressures on Panhandle just as the firm was heading into a complicated struggle for corporate survival.[70]

[69] Vietor, *Constrained Competition*, pp. 149–50. See pages 147–9 for background on Order 436. Also see FERC Order 436, Stats. and Regs. 30,665 (November 1985), 31,467–468; 50 Fed. Reg. 42,408 (October 18, 1985).

[70] The "New" Panhandle Eastern: A presentation to the New York Society of Security Analysts New York City, May 27, 1987. By 1987, Panhandle Eastern's rate structure remained unresolved. In February of that year, FERC issued Opinion 265 which overruled previous rulings by an administrative law judge which upheld the sole supplier provision of Panhandle Eastern Pipe Line's tariff. For the sole supplier tariff, Panhandle Eastern was obligated to supply the full requirements of certain gas customers, and these customers were obligated to buy gas only from PEPL. The FERC had preferred the sole supplier tariff because it believed that minimum bills distorted competition between gas sellers and resulted in higher overall gas costs. Opinion 265 found PEPL's sole supplier tariff to be anticompetitive. As a result, twenty-four of PEPL's customers contracted with the "sole supplier" tariff, were able to purchase natural gas on the spot market instead of from PEPL whereas PEPL remained obligated to honor its gas purchase contracts with producers. Subsequently, Panhandle Eastern accepted transportation requests from fifteen of its customers freed from the sole supplier requests. In another ruling, the FERC issued Order 451 during July 1986, which provided a mechanism allowing producers to receive higher prices for "old" gas. This was also a problem for Panhandle Eastern because 90 percent of its contracted domestic reserves were subject to this order. Order 451 also gave pipelines the opportunity to release large volumes of contracted production as a means of reducing future take-or-pay exposure. See FERC, Docket RM86-3, "Ceiling Prices: Old Gas Pricing Structure," Order 451, June 6, 1986.

The market for corporate control

By early 1986, Panhandle's board was devoting most of its attention to the very real threat of a takeover. In January, the Board adopted an extensive severance plan for senior executives and other employees that would be activated by a takeover. The following month, the Board invited eight representatives from Kidder, Peabody, Merrill Lynch Capital Markets, and the New York law firm of Sullivan & Cromwell to make a second formal presentation about appropriate defense mechanisms; the advisors recommended use of a so-called poison pill defense. The board agreed and quickly adopted a "Shareholder Rights Protection Plan." Under this plan, stockholders had the right, in the event of an attempted takeover, to buy additional stock of the surviving company at a 50 percent discount. They could thereby exact a heavy toll on the takeover organization.[71]

That same month in 1986, the company's takeover battle began. On March 12, 1986, Paul Taylor, director of investor relations, was attending an American Gas Association meeting in Boston when he received a phone call from Georgeson and Company's "Stockwatch" monitors. The monitors had noticed an unusually high level of trading in Panhandle stock. This news prompted the firm to launch a carefully crafted, high-profile public relations campaign at its fifty-seventh Annual Meeting in Houston on April 23, 1986. "Staying Power" was the motto.[72]

The company printed biographical synopses of six prominent Americans who had achieved great success, sometimes under adverse circumstances, in highly competitive situations. The six included former UCLA basketball coach John Wooden, actress Patricia Neal, author Isaac Asimov, jazz musician and composer Dave Brubeck, Penn State football coach Joe Paterno, and the "Peanuts" cartoonist Charles Schulz. The campaign included poster reprints with a photograph of each person and a text based on the subject's ideas about the concept of "staying power" in the midst of adversity. None of the persons spoke directly, or even indirectly for that matter, about Panhandle Eastern or the energy industry. The first "Staying Power" ad appeared in the *Wall Street Journal* on the morning of Panhandle Eastern's 1986 annual meeting, and the company ran one a

[71] Panhandle Eastern, *Minutes*, March 11, 1986.
[72] "Dear Fellow Employee . . . ," *Spectrum*, April/May, 1986, p. 3.

Figure 9.3 "Staying Power" advertisement (John Wooden)

week for six months in both the *Journal* and the *New York Times*.[73]

Popular as the campaign posters were, Panhandle's executives knew they needed a more material defense as well. They began confidential discussions with a possible friendly "suitor," Texas Eastern Corporation. Texas Eastern was a diversified energy firm with natural gas pipelines serving northeastern markets, profitable operations in the North Sea, a large retail LPG distribution system, and a large real estate development in downtown Houston. On May 2, Dick O'Shields and Robert Hunsucker had lunch with Texas Eastern's David Bufkin, chairman and CEO, and Dennis Hendrix, president and chief

[73] The ad campaign was the brain child of Bob Rehak, an advertising executive with the Houston office of Ogilvy and Mather. Rehak originally presented the concept during the summer of 1984 to Stan Wallace (public affairs) who liked the idea but did not feel the time was right. In the fall of 1985, Wallace and Rehak presented the idea to the Panhandle Eastern board. One week later, Rehak received approval to develop it. Over the next six months, Rehak settled on six persons and interviewed all of them at either their home or office. The internationally known photographer, Yousuf Karsh, who had taken the famous photograph of Winston Churchill for the cover of *LIFE* magazine in 1941, was hired to take the photographs. Bob Rehak, telephone interview by Christopher J. Castaneda, August 1993.

operating officer.[74] They discussed the possibility of combining the two companies, but "the relative stock prices and dividend rates of the two corporations made a statutory merger practically impossible at that point in time."[75]

Later that month, Panhandle's executives learned that an unidentified buyer had purchased a large number of PEC shares. After several days of research, they discovered that the new buyer was masked by an account managed by Bear, Stearns & Company Anticipating a tender offer, they began formulating a defense plan. During the next several weeks, no offer arrived.[76]

The mystery buyer emerged on June 20 when Panhandle's O'Shields received a phone call from William Harvin, a senior partner of the Houston law firm of Baker & Botts. Harvin said his client, Jack E. Brown, a partner in the Midland, Texas-based oil and gas partnership of Wagner & Brown, wanted to meet to discuss a possible "business combination." Later in the day, Brown and Harvin met with O'Shields, Hunsucker, and Panhandle's general counsel John Sieger. The meeting was inconclusive and there was no explicit discussion of a merger or acquisition. But Brown – who along with Wagner had been active in other takeover attempts – did inform the Panhandle group that Wagner & Brown had acquired Panhandle stock and indicated that T. Boone Pickens, Jr., of Amarillo was not involved in their efforts. Previously, the partners had joined an unsuccessful takeover attempt of Unocal Corp.[77]

At a subsequent meeting, Brown said that during the previous weeks he and his partner had acquired nearly five percent of Panhandle's stock (the maximum amount a firm could acquire without disclosure to the SEC). They bought the stock

---

[74] Dennis Hendrix joined Texas Gas Transmission Corporation in 1976 as its president and chief operating officer. He became CEO in 1978 and chairman of the board in 1982. Born in Selmer, Tennessee, in 1940, Hendrix received an MBA at Georgia State and then earned a CPA degree. He became president and CEO of Texas Eastern in November 1985. See "Offer to Purchase for Cash up to 48,650,000 Shares of Common Stock of TEC . . . by Pan Acquisition Co.," February 21, 1989, p. 15.

[75] Richard L. O'Shields to Christopher J. Castaneda, June 4, 1995.

[76] Irwin, *Summer of '86*, p. 20.

[77] "Wagner & Brown seeks to buy Panhandle Eastern," *Oil and Gas Journal*, July 7, 1986, p. 35.

through a partnership named Star Partners and now wanted 100 percent ownership of Panhandle Eastern.[78] Brown was prepared to offer $50 per share, $30 of which would be in cash and $20 of which would be in the form of nonredeemable preferred stock. Brown, who apparently knew that Panhandle's next board meeting would be in two days, expressed hope that the board would accept the $50 offer.[79]

Wagner & Brown's offer was all the more troublesome to Panhandle, because of Drexel Burnham Lambert's involvement as Wagner & Brown's financial advisor. Drexel had played a leading role in many of the nation's largest hostile takeovers. Two years earlier, Drexel backed an unsuccessful attempt by Mesa Petroleum, controlled by T. Boone Pickens, to buy Gulf Oil. Subsequently, Chevron purchased Gulf for $13.3 billion in a "white knight" rescue from the Mesa takeover attempt. Pickens's Mesa Petroleum had purchased Panhandle Eastern's former Hugoton Production Company in 1967, and this connection may have helped fuel unsubstantiated speculation that Pickens was involved in the Wagner & Brown bid for Panhandle Eastern. Drexel was well known for the talents of its "Junk Bond King," Michael Miliken, whose accomplishments had at this point earned him praise and great wealth.[80]

---

[78] "Wagner & Brown seeks to buy Panhandle Eastern," *Oil and Gas Journal*, July 7, 1986, 35; and Irwin, *Summer of '86*, p. 25.

[79] Brown noted that his investment advisors believed that there would be no problem in raising the required cash. Irwin asked Brown for the source of his financial advice. Brown replied that it was Drexel Burnham Lambert. Hunsucker closed the meeting soon thereafter, and he told Brown that he could not recommend to the board that the company be sold at that time. Brown later confirmed his offer in a letter to Hunsucker in which he suggested that Star Partners desired the expertise of current management. Apparently, Wagner & Brown hoped to win over management through an equity sharing plan. The letter stated that "Additionally, it is our plan to set aside a meaningful portion of the common equity of the new enterprise to be granted, at no charge, to key personnel as future employment incentives." Cyril Wagner, Jr., and Jack E. Brown to R. D. Hunsucker, June 23, 1986. Also see, Irwin, *Summer of '86*, p. 27–8.

[80] Connie Bruck, *The Predators' Ball: The Junk-Bond Raiders and the Man Who Staked Them* (New York: The American Lawyer/Simon and Schuster, 1988); and Miliken would later receive a jail sentence, as well. Dan G. Stone, *April Fools: An Insider's Account of the Rise and Collapse of Drexel Burnham* (New York: Donald I. Fine, Inc., 1990).

Hunsucker formally replied to Brown three days after their meeting. "We find it unlikely," he wrote, "that stockholder value could be maximized by a present sale of the company. . . . we do not intend to engage in discussions or negotiations with you at this time." Hunsucker, citing Star Partners's letter literally, noted that it was not clear whether Star Partners was making a firm offer or simply stating its intention to negotiate. The next day, Wagner & Brown clearly restated their intention to purchase Panhandle stock in a formal offer.[81]

To discourage erroneous stock market rumors, Panhandle Eastern issued a brief press release disclosing as little as possible while trying to avoid accusations that the company had not been forthright about an offer. The release stated:

> Panhandle Eastern Corporation (PEC), a diversified energy company, announced today that it has received a request from a third party to negotiate the possible acquisition of the Company. The third party was informed that management did not intend to engage in discussions or negotiations of an acquisition at this time.

Stockholders and arbitragers, anxious to learn whatever they could about Panhandle's plans, inundated the firm with inquiries.[82]

Settling with Sonatrach

It was evident to Hunsucker and his team of executives that survival depended upon raising Panhandle's stock price, making acquisition more expensive. This might be accomplished by spinning off Anadarko, management reasoned, but such a move could not be undertaken while the firm was engaged in the corporate life threatening LNG litigation with Sonatrach. They decided to call the firm's $200 million bank credit and then offer that sum to the Algerians in cash, plus six million shares of Panhandle Eastern stock. This would be over 10 percent of the firm's outstanding stock, and if it was valued at $50 per share, they would be offering a total settlement of $500

---

[81] R. D. Hunsucker to Cyril Wagner, Jr., and Jack E. Brown, June 26, 1986 and Cyril Wagner, Jr., and Jack E. Brown to R. D. Hunsucker, June 27, 1986.

[82] Panhandle Eastern, *Press Release*, June 1986; and "Pickens silent on bid for Panhandle Eastern," *Amarillo Daily News*, June 28, 1986, p. 9-A.

million. Then they would issue $200 million in new long-term debt to cover the $200 million revolving credit loan for the cash payment.[83]

The Panhandle Eastern delegation met with Dr. Youcef Y. Yousfi, Director General of Sonatrach, in Geneva at the Hotel Des Bergeres. They presented the $500 million proposal and told Yousfi that if the takeover of Panhandle was successful, it was doubtful that Sonatrach's claims against Panhandle Eastern could ever be paid, even if Sonatrach eventually won the arbitration. Yousfi wanted an analysis of the situation by his enterprise's advisors, Dillon, Read & Co. Pending their evaluation, Yousfi indicated that he would recommend acceptance of the settlement to the Algerian government. Panhandle set a time limit of July 9 for official acceptance of the offer.[84]

At Panhandle's board meeting on July 8, the directors officially rejected Star Partners bid, referring in particular to the "speculative nature of the preferred stock."[85] The board also confirmed management's desire to continue settlement negotiations with Sonatrach. In the meantime, the firm would continue negotiating with Wagner & Brown, if only to allay any suspicions on their part that Panhandle Eastern was developing an alternative plan. Panhandle would also seek to buy out its Lachmar partners and escape that additional liability.

The first part of the puzzle to fall into place was the Lachmar contract. In the same afternoon of the Wagner & Brown meeting, a Panhandle working group flew to New York. They met with Wayne Welles, General Dynamic's Chief Financial Officer, and Paul Tregurtha, President and CEO of Moore–McCormack Resources Corporation. Welles and Tregurtha agreed to let Panhandle Eastern buy out their firms' interest in Lachmar for a total of $135 million.[86]

Next came a formal agreement with Sonatrach. The two groups met in Paris while the lawyers were busy preparing the French language version of the settlement agreements that would accompany the English language contract. Yousfi reported, however, that he had problems in explaining to Algerian authorities why Panhandle Eastern's settlement was substan-

[83] Irwin, *Summer of '86*, p. 44.
[84] Ibid., 67.
[85] Panhandle Eastern, *Minutes*, July 8, 1986.
[86] Panhandle Eastern, *Annual Report, 1987*, p. 51.

tially less than what Sonatrach hoped to gain through arbitration. The meeting ended on this note and the Panhandle executives went back to their hotel rooms to fret over the deal. Late that night, they met again with the Sonatrach delegation. By 2:00 A.M., Yousfi still found the proposed agreement problematical. Even after Hunsucker and Yousfi took "a walk in the garden," they could not agree.

Panhandle's representatives decided that they could not offer more cash, or more than the six million shares of stock approved by the board. Hunsucker and Yousfi then took another walk. After more than another hour, they finally returned. Hunsucker had agreed to extend the stock repurchase offer for a second year, at $55 per share. At 5:00 A.M. after discussing the deal with Algerian authorities, Yousfi reported that he could accept the settlement. Panhandle had paid a very high price, but it was now free of the LNG deal.[87]

Hunsucker later commented that the Wagner & Brown takeover bid was "probably one of the best things that ever happened to us. It brought things to a head." Indeed, the offer forced Panhandle management to make the tough decisions necessary to settle the LNG arbitrations, and it also forced the Algerians to settle with Panhandle rather than take their chances with Wagner & Brown and its risky financial program.[88] In this and many other ways, the market for corporate control forced American management to confront reality in the 1970s and 1980s.

Panhandle Eastern representatives then confronted the task of explaining the accounting effects of the LNG settlement to security rating agencies. In the meantime, the price of Panhandle stock, which had hit $50 per share after the Wagner & Brown offer, fell to the low $40 range by August 1. Stockholders, particularly takeover arbitragers, were unhappy that Panhandle was rejecting a $50 offer while apparently doing nothing to provide stockholders with maximum share value. These concerns proved to be ill-founded.

Panhandle's managers moved forward with their plan to spin

[87] "Panhandle to settle LNG import claims," *Oil and Gas Journal*, July 28, 1986, p. 40.

[88] Robert D. Hunsucker, interview by Christopher J. Castaneda and Clarance M. Smith, October 1, 1993, PECA. Also see, "How Panhandle Disarmed a Time Bomb," *Business Week*, August 4, 1986, p. 29.

off Anadarko to their stockholders. There were numerous legal and accounting ingredients to this strategy, but PEC gradually surmounted the difficulties. During the fall of 1986, Anadarko became an independent firm with $1.3 billion in assets, 20.8 million barrels of oil revenues, 1.6 Tcf of natural gas reserves, and over 650 employees. Anadarko had an interest in 3,039 producing oil wells and 2,131 producing U.S. natural gas wells, most of which were located in the Kansas portion of the Hugoton field.[89] Immediately following the spin-off, Anadarko stock traded on the New York Stock Exchange at $19.50 and Panhandle stock hovered between $24.25 and $28.125. By the summer of 1987, the combined Panhandle and Anadarko price hit $52.125 per share. At the end of that year, the aggregate price reached nearly $70 per share.

The spin-off arrangement did not meet with everyone's approval. Very soon after the transaction, Anadarko sued Panhandle Eastern in regard to six issues related to the spin-off agreement. One of these involved Panhandle Eastern's insistence that Anadarko continue to dedicate valuable Matagorda Island gas reserves to Panhandle Eastern. "They forced us to sign a contract," recalled Anadarko's Robert Allison, "that we would guarantee to sell that to them at whatever price they wanted to pay and they would take whatever they wanted whenever they wanted."[90] Panhandle was unable to control the reserves it once owned, but it had retained its independence.

Nevertheless, the firm's supplemental fuel project write-offs – particularly LNG and coal – scarred the company's financial statement. For the year, the corporation suffered a consolidated after tax loss in net income of $542 million, or $11.06 per share (Table 9.2). The after tax charge of the LNG settlement was $468 million, the Y&O Coal Company write-down was $45 million, and the Dixilyn-Field lost another $122 million.[91]

---

[89] "Panhandle to Spin Off Anadarko," *Oil and Gas Journal*, August 25, 1986, p. 34.

[90] *AnadarkoPetroleum v. Panhandle Eastern*, Delaware Supreme Court. 545 A.2d 1171 (1988). Also see John A. Sieger, interview by Christopher J. Castaneda and Clarance M. Smith, December 21, 1993. Robert J. Allison Jr., Paul A. Taylor, and Charles G. Manley, interview by Christopher J. Castaneda and Clarance M. Smith, December 20, 1993, PECA. Irwin, *Summer of '86*, p. 110.

[91] The Dixilyn properties were later sold. Grace Drilling, the largest operator of land-based rigs in the U.S., purchased the twenty-two land rigs for stock, and

Table 9.2. *PEC consolidated financial data, 1978–1988*

Year	Operating revenues	Net income	Earnings per share[a]	Long-term debt	Total assets
1978	$1,409	$117	$3.29	$723	$2,218
1979	1,966	146	4.01	821	2,648
1980	2,469	182	4.86	1,159	3,235
1981	2,870	262	6.52	1,104	3,571
1982	3,015	220	5.36	1,473	4,379
1983	3,145	152	3.64	1,438	4,454
1984	2,904	167	3.88	1,403	4,536
1985	2,503	125	2.83	1,156	4,170
1986	2,059	(542)	(11.06)	1,092	3,025
1987	1,465	108	2.04	1,033	3,045
1988	1,156	(157)	(2.86)	1,206	2,973

[a]Earnings (loss) per common share from net income (loss).
Source: Panhandle Eastern Corporation, *Financial and Statistical Supplement to the 1991 Annual Report* and Panhandle Eastern, *Annual Report*, 1983, 1993.

Panhandle Eastern's total assets then amounted to $3 billion; they included Panhandle Eastern Pipe Line Co., Trunkline Gas Co., Dixilyn–Field Drilling Co., two coal companies, gas lines located in Kansas, and Century Refining Company.

Getting Panhandle back on course

Having escaped the burdensome LNG contract, Panhandle was able to reenter that business under more favorable conditions. On April 26, 1987, the company reached a new agreement with Sonatrach to market regasified LNG in the U.S. on a "market sensitive" basis. Sonatrach made available 3.3 Tcf of reserves from the Hassi–R'Mel Field, with no minimum, take-or-pay volumes.[92] Trunkline LNG and Citrus Trading Corporation then

Sonat Offshore Drilling Inc. purchased the ten offshore rigs for a fixed percentage of future net cash flow.

[92] The agreement gave Sonatrach until August 10, 1988 to exercise the put for a total of $330 million. In further negotiations between Panhandle Eastern and Sonatrach, the two firms agreed to cancel the put option and Panhandle Eastern issued Sonatrach an additional 1.5 million shares. All issues arising from the LNG arbitration were finally settled. Panhandle Eastern, *Minutes*, April 27, 1988; July 27, 1988; August 9, 1988.

contracted for 120 MMcf/d to be delivered to Citrus through Florida Gas Transmission.[93]

Panhandle's management also cleared up the remaining debt from the LNG tanker deal, after a series of contentious negotiations and a lawsuit. PEC now owned the two LNG tankers entirely and also bore complete responsibility for paying the debt service on the ships. The best way out of this bind, Panhandle decided, was to default on the bonds. In November 1987, it declined an interest payment on the *S.S. Lake Charles*; in March 1988 it declined payment on the *S.S. Louisiana* debt as well. As a result, the U.S. government foreclosed on the bonds and took control of the tankers.[94] Subsequently, the U.S. District Court of Delaware returned a summary judgment against Trunkline and TLC for $147 million, approximately the balance of the outstanding debt on the tankers, plus accumulated interest. Panhandle Eastern, after losing an appeal, paid the judgment and reacquired ownership of the tankers.[95]

Meanwhile, the FERC helped Panhandle and other pipelines deal with their take-or-pay liabilities. The agency's new Order 500 allowed the pipelines to recover half of the loss through a rate surcharge and to include one-quarter in its commodity rates.[96] The pipelines still had to absorb the loss, but now the utilities could no longer unilaterally cancel the supply of gas for which they had contracted.[97] The relief this provided was significant. Through 1988, the Panhandle Eastern gas system paid $694 million to extinguish most of a $4.3 billion take-or-pay exposure. The Trunkline Emergency Marketing Program and spot market sales of released volumes helped Trunkline make the transition through this phase of its corporate crisis.[98]

---

[93] Florida Gas Transmission was a subsidiary of Enron Corporation.

[94] Tom Irwin, "The MARAD Bond Default," April 20, 1994.

[95] Panhandle Eastern, *Minutes*, September 20, 1988.

[96] Vietor, *Contrived Competition*, p. 158; Castaneda and Pratt, *From Texas to the East*, p. 252.

[97] FERC Order 500, Fed. Reg. vol. 52, no. 157 (August 14, 1987), pp. 30334–57. Richard Vietor greeted Order 500 with the statement: "When the court affirmed this rule in 1988, regulated competition became an established fact," in *Contrived Competition*, p. 161. Vietor may well have spoken too soon. In the next decade, one more FERC Order, Order 636, firmly and resolutely imposed a system of regulated competition which superseded the rules of Order 500.

[98] Panhandle Eastern, *Annual Report, 1988*, p. 11.

Figure 9.4 Panhandle Eastern Corporation headquarters complex in Houston, Texas

Like many other U.S. firms during these years, Panhandle cut costs and reduced its staff. By the mid-1980s, the firm's staff numbered approximately 1,100 and occupied three buildings in Kansas City and one in Houston. By consolidating all of the administration in Houston, the company saved money and eliminated a great deal of travel between the two cities.[99] There were painful cuts – the overall work force was reduced by 20 percent – and cultural tensions arising from the firm's consolidated work force. But by November 1988, Panhandle had substantially completed the relocation and restructuring of its central administrative offices in Houston.[100]

By that time, it was apparent that Panhandle Eastern would survive its most recent corporate crisis. In this case as in others, the market for corporate control had accelerated the transition to a more competitive, efficient, and innovative style of organization. Encouraged by the regulatory policy of the postwar era and the shortages that policy induced, Panhandle's managers aggressively sought to capitalize on supplemental

[99] Herbert E. Schulze, interview by Christopher J. Castaneda and Mack Price, August 9, 1991, PECA.
[100] Dan Hennig, interview by Clarance M. Smith, November 22, 1991, PECA.

fuel ventures. In the process, however, they overextended the firm and incurred enormous liabilities in a failed attempt to exploit LNG, a new energy source, and the technology it required. No longer constrained as they had been in the past by William Maguire's personal, and conservative, style of management, they bet the company's future on a single major innovation and almost lost the firm. Under tremendous financial pressure, they cleared the company's major debts, cut its costs of operation, and enhanced stockholder value as well.

They did successfully steer the firm through the process of deregulation. That process was continuing, but by the late 1980s, it was already evident that markets and not regulators would be the major force shaping the natural gas industry in future years. What was not yet clear was the role pipelines would play in the industry. The trend away from vertically integrated firms that bought, transported, and sold gas on a national and international basis was pronounced. Panhandle, like its competitors, was devoting more of its resources to transportation, less to the merchant gas business. But it remained to be seen whether this industry would be dominated by segmented, single-function companies or by the type of vertically integrated corporations characteristic of most other sectors of the American economy.

10

The new order

While Panhandle's executives had good cause to be pleased with what they had accomplished by the late 1980s, they had little opportunity to reflect on the organization's progress. For one thing, the LNG and take-or-pay settlements had left the firm's capitalization level as high as it had ever been. New regulations designed to transform pipelines from natural gas merchants to transportation services were also threatening to restructure an industry historically accustomed to buying and selling gas. Moreover, Panhandle Eastern's midwestern market area was no longer a growth market. Panhandle had survived an incredibly complex ordeal, but its future was still very uncertain.

CEO Hunsucker, his fellow executives, and the Board decided to deal with this situation by opting for growth rather than retrenchment: they first set out to seek new gas markets. In early 1988, the firm approved a plan to form the Indiana–Ohio Pipeline Company to connect with Texas Eastern's system in order "to facilitate the transportation of southwestern gas to east coast markets."[1] This proposal brought a quick competitive response from American Natural Resources (ANR). ANR included the former pipeline subsidiaries – Mich–Wisc

[1] Panhandle Eastern, *Minutes*, February 2, 1988, p. 10.

and American Louisiana – of the American Natural Gas Company (the former parent of MichCon) and comprised a similar and competitive midwestern pipeline system. Now, under the control of the Coastal Corporation and its chairman Oscar Wyatt, ANR, acted under the FERC's Northeast "open season" settlement and proposed a competing ninety-one-mile pipeline from Muncie, Indiana to Lebanon, Ohio.[2]

Acquisition

Then, on January 16, 1989, Coastal mounted a far more serious challenge to Panhandle's growth strategy, and forced Panhandle Eastern to quickly examine its own long-term plans for increasing its throughput and expanding into new markets. Coastal offered $42 per share, for a total of $2.6 billion, for Texas Eastern Corporation. Texas Eastern's 10,567 mile system was one of only three major pipelines extending from the southwest into northeastern markets. Texas Eastern's access to that large energy market – including New England, through its Algonquin Gas Transmission system – made it a very attractive acquisition.

This was especially true where Coastal – Panhandle's primary competitor in the Midwest – was concerned. Originally founded as Coastal States Gas Producing Company in 1955 by Oscar Wyatt, Jr., Coastal began business as a gas gathering firm. In the early 1970s, Wyatt had begun expanding by acquisition; first he had absorbed Colorado Interstate Corporation, a pipeline extending from the Texas Panhandle into Colorado and Wyoming. In 1985, Coastal acquired American Natural Resources for $2.5 billion. This move transformed the Coastal system into a major player. Wyatt and Texas Eastern's CEO, Dennis Hendrix, had already struggled in the early 1980s through one hostile takeover attempt when Hendrix was president and CEO of Texas Gas Transmission (the firm which acquired Mo–Kan's original Kentucky Natural Gas properties).

---

[2] Panhandle Eastern, *Minutes*, February 1, 1989, p. 18. American Natural Gas became American Natural Resources (ANR) on May 2, 1976. Also see George Mazanec, interview by Christopher J. Castaneda and Clarance M. Smith, January 14, 1994, PECA, for a review of the "Lebanon Lateral" story.

Hendrix was not looking forward to another round of corporate combat with the aggressive Wyatt.[3]

Hunsucker learned about the attempt directly from Wyatt (shortly before the offer hit the market headlines), and Panhandle's CEO telephoned Hendrix, that same day to offer his help.[4] Later that week, Panhandle Eastern's forty-member management group considered a possible competitive tender offer for Texas Eastern. While the two CEOs continued their conversations, Panhandle officials looked into possible financing arrangements with investment bankers, spoke to other pipeline firms that might be interested in three-way consolidations, and discussed with other parties the possibility of dividing up Texas Eastern's assets.[5]

Panhandle's management group agreed that the company needed to expand its operations, that acquisition was the best strategy, and that Texas Eastern would provide access to a growing market. Official meetings between representatives of the two companies began the following week. On January 25, the two firms signed a confidentiality and ninety-day "standstill" agreement. Subsequently, they worked out the major elements of a possible tender offer with representatives from Kidder, Peabody, and Panhandle Eastern's legal advisors, and top executives from Texas Eastern.[6] Excitement and work levels built up tremendously. Hunsucker recalled: "I don't think I had, or any of us had, a day off for the next 30 days maybe longer, 7 days a week and all night. . . . " and they kept "quiet all the way through."[7]

Texas Eastern's management investigated alternatives to being acquired, including the possibility of a leveraged buy-

---

[3] For a review of the Texas Eastern takeover battle, see Castaneda and Pratt, *From Texas to the East*, pp. 256–7.

[4] Hunsucker, Robert D. Interview by Christopher J. Castaneda and Joseph A. Pratt, January 23, 1990, PECA.

[5] Robert D. Hunsucker, interview by Christopher J. Castaneda and Joseph A. Pratt, January 23, 1990, PECA. Also see James W. Hart, Jr., interview by Clarance M. Smith and Christopher J. Castaneda, December 17, 1993, PECA.

[6] "Offer to Purchase for Cash Up to 48,650,000 Shares of Common Stock of Texas Eastern Corporation at $53.00 Net Per Share by Pan Acquisition Co. a wholly owned subsidiary of Panhandle Eastern Corporation," February 21, 1989, PECA.

[7] Robert D. Hunsucker, interview by Christopher J. Castaneda and Joseph A. Pratt, January 23, 1990, PECA.

out.[8] They set up a data room open to any firm with a serious potential interest. Panhandle's working group "would go in the side door and they would take them up the elevator so no one would see them and were able to do that over I guess more than a 2 week period. It never leaked in any of the Texas Eastern offices or here. . . . " Hunsucker said it was "critical" to keep our "powder dry until we are ready to light the fuse and walk away from it." The possibility that Panhandle Eastern might be preparing to bid on Texas Eastern never appeared in the media.[9]

Texas Eastern's executives had quietly decided that Coastal's offer was not particularly good, and Wall Street agreed. Texas Eastern's stock continued to trade in the $47–50 range, reinforcing the firm's desire to look elsewhere for a buyer. Of course, Coastal's position was that its offer adequately reflected the business's real value. James R. Paul, Coastal's president and chief spokesman during its tender offer, wrote to Dennis Hendrix, "In the last two days, uninformed arbitragers and speculators have pushed the price of the Texas Eastern stock to levels which we believe, based on the public information presently available to us, are not supported by the fundamentals of your businesses."[10] Price was the key to the struggle.

On Sunday, January 22, James Paul and other representatives from Coastal met Texas Eastern's board and officers. The Coastal officials presented their company's reasons for the offer to "more than a dozen advisers alongside the directors in a cramped, standing room only board room."[11] Following the presentation, the Coastal representatives left, and Texas Eastern's board privately discussed the offer. Shortly thereafter, Hendrix met again with Paul and told him that "Texas Eastern didn't see any reason to continue the meeting."[12]

Texas Eastern's official response to Coastal's bid came on

[8] James B. Hipple, interview by Christopher J. Castaneda and Clarance M. Smith, December 22, 1993, PECA.

[9] Robert D. Hunsucker, interview by Christopher J. Castaneda and Joseph A. Pratt, January 23, 1990, PECA.

[10] Barbara Shook, "Coastal, Texas Eastern Execs to Meet," *Houston Chronicle*, January 19, 1989, p. B-1.

[11] Caleb Solomon and Dianna Solis, "Texas Eastern To Seek Bidders Against Coastal," *The Wall Street Journal*, January 23, 1989.

[12] Ibid.

Sunday, January 29, when its board "unanimously rejected as inadequate the $42 per share tender offer. . . . "[13] Hendrix reported that Coastal might raise its bid in response to a higher offer but it had refused "to allow a reasonable time for a fair bidding process to be conducted." Texas Eastern set its own deadline for a higher bid at March 15 – six weeks later.

Panhandle's directors held a special meeting on Thursday, February 15, 1989. Noting that Coastal had extended its original tender offer from February 14 to February 23, they officially considered a competitive offer for Texas Eastern's total assets valued at $5.45 billion (or $2.6 billion after discounting for liabilities). The directors agreed to meet again on Monday, February 19 to make a final decision as to whether they would extend an offer the following day. At their President's Day meeting they agreed to offer $52 per share for up to 80 percent of Texas Eastern's stock and to exchange Panhandle Eastern stock for the remaining 20 percent of Texas Eastern common. Later in the evening, after a meeting between representatives of the two corporations, Panhandle raised the price to $53.00, for a total of $3.22 billion. If successful, this would be the most expensive pipeline acquisition in U.S. history (Table 10.1). Indeed, a combined Texas Eastern–Panhandle network would resemble the Columbia Gas–Panhandle system of the 1930s broken apart by federal antitrust policy.

The $53 price, which was near the high range ascribed to the stock by Kidder, Peabody, matched the total value of Texas Eastern's assets. By offering top dollar, Panhandle hoped to avoid a bidding war, and it announced its offer that same day.[14] Since the stock market was closed for the federal holiday, Panhandle officials conducted discussions with analysts during the day in preparation for filing SEC documents on February 21. Hunsucker organized a merger committee and James W. Hart,

---

[13] Texas Eastern Corporation, News Release, January 29, 1989.

[14] Panhandle Eastern, *Board of Directors Minutes*, February 15, 1989; February 19, 1989. The precise exchange formula was each Texas Eastern share would be exchanged for not less than 2.038, nor more than 2.304 shares of Panhandle Eastern common stock, and the exact exchange ratio would be determined by dividing $53 by the average trading price per share of Panhandle Eastern common stock on the ten trading days immediately preceding the effective date of merger.

head of public affairs, quietly launched a media campaign.[15] This included a video reporting service to inform field employees, an Acquisition Hotline for employees, and an Acquisition Newsletter. Panhandle's consultants recommended that Panhandle Eastern specifically use the phrase "acquisition" to prevent any misunderstanding about what was taking place.[16]

Texas Eastern accepted the offer on April 27, 1989. To fund the acquisition it borrowed $2.6 billion from banks and sold $300 million in senior subordinated reset debentures. Initially, Panhandle acquired 81 percent of the outstanding common stock for $53 a share in cash. Later, it exchanged 27.1 million shares of its common stock for the remaining 19 percent of Texas Eastern's shares; for this transaction, Panhandle's common stock was valued at $23.75 per share.[17] Panhandle's 1989 Annual Report correctly described the deal as "the largest business development in the Company's 60-year history. . . . "[18]

The results of the acquisition provide an excellent example of how the market for corporate control worked to strengthen the American business system and indeed the entire economy. The resulting process was socially painful, but it resulted in a more efficient business entity focused more intensely on its core enterprises in natural gas. As a consequence of the combination, Panhandle was able to eliminate hundreds of overlapping jobs and to sell most of Texas Eastern's nonpipeline assets; early retirements and work force reductions followed.[19]

Soon after the bid, Panhandle announced its intention to sell Texas Eastern's North Sea interests, its Houston Center real estate development, Petrolane, Inc., and the La Gloria refinery.[20] Panhandle could then use the proceeds to help pay for its $3.22 billion tender offer. The Federal Trade Commission also

---

[15] James W. Hart, Jr., interview by Clarance M. Smith and Christopher J. Castaneda, December 17, 1993, PECA.

[16] Ibid.

[17] Panhandle Eastern, *Board Minutes*, June 28, 1989; Panhandle Eastern, *Annual Report, 1989*, p. 43; and Panhandle Eastern, *Annual Report, 1988*, p. 1 (insert).

[18] Panhandle Eastern, *Annual Report, 1989*, p. 11.

[19] Barbara Shook, "Panhandle Eastern to Cut Jobs," *Houston Chronicle*, April 27, 1989. Also see PEC–TEC, *Consolidation Policies and Benefits Program*, no date.

[20] Castaneda and Pratt, *From Texas to the East*, pp. 259–61.

Table 10.1. *Mergers and acquisitions of natural gas pipelines, 1982–1989*

Year	Acquiring company	Acquired company	Acquisition cost (millions)
1982	Northwest Energy	Cities Service Gas Co.	371.4
1983	Goodyear Tire & Rubber Co.	Celeron Corp.	828.0
	CSX Corp.	Texas Gas Resources Corp.	1,054.0
	Midcon Corp.	Mississippi River Transmission Corp.	256.3
	Burlington Northern, Inc.	El Paso Co.	1,278.0
	Williams Companies	Northwest Energy Co.	886.0
	Texas Oil & Gas Corp./MidCon Corp.	Tatham Pipeline Co.	200.0
1984	Cabot Corp.	Wester Transmission (Pioneer Corp.)	360.0
	Houston Natural Gas Corp.	Transwestern Pipeline Co.	390.0
	Houston Natural Gas Corp.	Florida Gas Transmission Co.	800.0

Year	Acquirer	Acquired	Value
1985	Coastal Corp.	American Natural Resources Co.	2,451.9
	Internorth, Inc. (Enron)	Houston Natural Gas Corp.	2,400.0
	Seagull Energy Corp.	ENSTAR Natural Gas Co.	65.0
	Tenneco, Inc.	Celeron Corp. (natural gas operations)	457.0
	MidCon Corp.	United Energy Resources, Inc.	1,160.9
1986	Occidental Petroleum Corp.	Midcon Corp.	2,608.0
	Arkla, Inc.	Mississippi River Transmission Corp.	305.0
	Hadson Petroleum Corp.	Llano, Inc.	41.4
	Oklahoma Gas and Electric Co.	Mustang Fuel Corp.	120.6
	Sonat, Inc.	Florida Gas Transmission Co. (50%)	360.0
	Texas Eastern Corp.	Algonquin Energy, Inc.	117.4
	Pacific Gas & Electric Co.	Pacific Gas Transmission	164.0
1987	LaSalle Energy Corp.	United Gas Pipeline Co.	460.0
1989	Panhandle Eastern Corp.	Texas Eastern Corp.	3,220.0

Source: Arthur Andersen & Co., Cambridge Energy Research Associates, Natural Gas Trends (Cambridge, Mass. 1989), and Vietor, Contrived Competition, pp. 348–9.

Table 10.2. *Texas Eastern asset sales, 1989*

Date of sale (completed)	Property	Purchaser	Amount (millions)
April, 1989	TE Norwegian North Sea	Enterprise Oil PLC	$ 408
April, 1989	Eastman Christensen (50%)	Norton Company	115
May, 1989	Indonesian reserves	Enterprise PLC	13
August, 1989	Petrolane MLP (44%)	Quantum Chemical Corp.	450
October, 1989	La Gloria Oil & Gas Co.	Crown Central Petroleum	110
September, 1989	TE UK North Sea	Enterprise Oil PLC	961
November, 1989	TE Products Pipeline Co.	MLP (88.25%)[a]	588
December, 1989	Houston Center	JMB Realty	400
Total			$3,045

[a]Master Limited Partnership in which PEC retained operating control.
Source: Panhandle Eastern, *Annual Report, 1989,* p. 43.

mandated PEC to divest its interest in Stingray, since it over-lapped part of Texas Eastern's offshore system. In 1991, PEC sold Stingray for $32.5 million. Panhandle was able to sell most of Texas Eastern's subsidiaries well before the end of the year. On March 1, 1989, the firm sold the most valuable non-pipeline asset, the North Sea oil reserves. Sales of the large retail LPG operations, oil refinery, and real estate holdings followed quickly. By the end of 1989, in fact, there were only $53.7 million in assets still up for sale (Table 10.2).[21]

[21] "Enterprise Leaps Into North Sea Big League," *Financial Times,* March 2, 1989, p. 17. Also, Barbara Shook, "Texas Eastern Plans $1.4 Billion Sale," *Houston Chronicle,* p. B-1. Before the deal was completed, British Gas PLC of London and Amerada Hess Corp. of New York, both of which were partners with Texas Eastern in the original seismic testing and development of the North Sea in the early 1960s, indicated that they wanted to purchase Texas Eastern's North Sea properties. When the original group agreed to conduct seismic tests jointly and then develop the North Sea, they had granted each other the right of first refusal on any sale of their interest in the North Sea. Eventually, British Courts ruled that Enterprise Oil, which had been spun off by British Gas years earlier, could legally purchase the North Sea proper-ties, and that deal was consummated. Anne Pearson, "Texas Eastern Agrees to Sell Petrolane Inc.," Houston Chronicle, May 31, 1989, p. B-1. Also, Ran-

From these transactions, Panhandle recovered gross revenues of approximately three billion dollars. Texas Eastern's high break-up value had virtually ensured its ultimate demise as a separate firm, and now the sell-off allowed Panhandle to pay down its $3.22 billion takeover debt. Proceeds from these sales and other sources reduced the bank acquisition debt from $2.6 billion to $424.6 million at the end of the year. The company's debt-to-capitalization ratio dropped as a result from 80 to 68 percent.[22] Panhandle quickly announced plans for a 225 MMcf/d expansion of its northeastern system in order to utilize more fully the capacity of the Trunkline system.

Management succession

There was considerable speculation regarding the next leader of the new Panhandle Eastern Corporation. Robert Hunsucker, who would soon reach the mandatory retirement age of sixty-five, had to choose a successor. Some press reports speculated that Texas Eastern's Hendrix would get the top job.[23] Hunsucker, with the help of a search firm, opted for Philip J. Burguieres. A Panhandle Eastern director since 1984, Burguieres had most recently been CEO of the Cameron Iron Works, Inc. He had successfully revived the steel firm's financial situation, and he soon had an opportunity to do the same for Panhandle. His immediate goals included reducing debt and raising earnings in order to pay Panhandle Eastern's $2.00 per share dividend.[24]

dall Smith, "Panhandle Unit is Set for Sale at $1.18 Billion," *Wall Street Journal*, May 30, 1989, p. 5. Ralph Bivins, "Houston Center Sold for Whopping Price Tag," *Houston Chronicle*, September 16, 1989, p. 1. Anne Pearson, "Panhandle Plans Pipeline Sale," *Houston Chronicle*, November 23, 1989, p. D-1.

22 PEC, *Annual Report, 1989*, p. 5.

23 Caleb Solomon, "Texas Eastern Chief Caught Amid Suitors," *Wall Street Journal*, March 10, 1989.

24 Anne Pearson, "Panhandle Eastern Elects New Chief," *Houston Chronicle*, November 30, 1989. Of Texas Eastern's officers, only George Mazanec, senior vice-president of gas operations; James B. Hipple, vice-president and chief financial officer; and Paul Ferguson, treasurer, were retained by Panhandle Eastern. Mazanec replaced Richard Dixon who was retiring as one of Panhandle Eastern's group vice-presidents. Previously, Mazanec had worked for Hendrix at Texas Gas in the early 1980s. Hipple became senior vice-president, finance, and Paul F. Ferguson, Jr., became treasurer.

Burguieres was immediately confronted by a series of difficult managerial problems. In an effort to improve the efficiency of the organization, he introduced a Total Quality Management program of the sort popular throughout American business during these years. He attempted to upgrade Panhandle's accounts receivable system and brought in a consulting management group to assist in developing a new organizational structure.[25]

But during the summer of 1990, Panhandle's common stock fell below $20 per share; it had traded at $30 in late 1989 and early 1990. The New York Stock Exchange asked for a statement explaining the decline in the stock price, and Panhandle announced that it would review its current dividend policy at the board's upcoming (July 25) meeting.[26] The renewed LNG sales were sagging; Texas Eastern's favorable Gulf Warranty contract (long-term, low-price) had expired; and its polychlorinated biphenyls (PCB) liability weakened its financial condition. It appeared that Panhandle Eastern, which had paid dividends continuously since 1939, would not be able to earn its current dividend through revenues.[27]

Management and the board decided to lower the quarterly dividend from $.50 to $.20 per share effective September 15, 1990.[28] This increased the downward pressure on the stock, which fell to the $10 range by late October. The dramatic drop attracted the attention of Wall Street speculators and analysts. On Tuesday, October 30, Dan Dorfman, CNBC financial correspondent, reported that New York broker Alan Gaines held for clients a short position of 750,000 shares of Panhandle stock and would benefit from an even lower price. Dorfman said that

---

[25] James W. Hart, Jr., interview by Clarance M. Smith and Christopher J. Castaneda, December 17, 1993, PECA.

[26] Panhandle Eastern, News Release, July 5, 1990.

[27] James B. Hipple, interview by Christopher J. Castaneda and Clarance M. Smith, December 22, 1993, PECA. For a discussion of Texas Eastern's PCB problems see, Castaneda and Pratt, *From Texas to the East*, pp. 253, 258.

[28] Panhandle Eastern, *Minutes*, July 25, 1990. Chief Financial Officer Charles Lasseter opposed the cut, and Panhandle announced his early retirement the day before publicly announcing the dividend cut. Burguieres had earlier dismissed his General Counsel, John Sieger, over disagreements on staffing the legal department; Sieger was replaced by Carl B. King, who had served as president of Cameron's Oil Tools Division and previously as General Counsel at Cameron Iron Works. See "Philip Burguieres to the Stockholders," July 25, 1990, PECA.

"his [Gaines's] view is the stock, in the first quarter, will be around six bucks a share. And he thinks there's a very good shot it may go bankrupt. . . . " Dorfman, who had telephoned Panhandle Eastern the night before, ridiculed the company's refutation of Gaines's statements.[29]

The next day, after receiving a letter from Panhandle briefly summarizing positive developments in the company's financial structure, Dorfman noted that "Standard & Poor recently upgraded the debt to investment and Panhandle also has three hundred million in cash. Panhandle is essentially saying that the short seller is all wet."[30] Nevertheless, it appeared at this time that the same market that had reshaped the firm might be about to sink it. To add to the problems, CEO Burguieres, under intense pressure, began to suffer serious health problems. No longer able to serve as CEO, Burguieres told the board to look for a replacement.[31]

An obvious candidate was close at hand, and the board quickly offered the position to Dennis R. Hendrix, the former CEO of Texas Eastern.[32] During negotiations, Hendrix agreed to become the new chairman, CEO, and president of Panhandle Eastern. Hendrix recalled:

> When I was approached, it was apparent that they needed someone to come in. I had some familiarity with the company, had some background with the company, and had a natural vested interest in terms of having financial involvement in the company as well as a lot of friends that were involved. So, it didn't really take a lot of time for me to weigh the circumstances and say, 'Yeah, I'll come in and do what I can to try to deal with the particular problems and challenges that the company was confronted with' at that time. I didn't have a lot of time to think about it quite honestly, because it was presented on a very short fuse. They needed some action taken.[33]

---

[29] See Transcript of *Business View*, CNBC-TV, October 29, 1990: 3:30–4:30 P.M.; *Money Wheel*, CNBC, October 30, 1990: 11:00 A.M.; John Barnett to "News in Review" Distribution, October 31, 1990; and James W. Hart, Jr., to Dan Dorfman, October 30, 1990.

[30] Ibid.

[31] Panhandle Eastern, *Minutes*, November 7, 1990.

[32] Ibid.

[33] Dennis R. Hendrix, interview by Christopher J. Castaneda and Clarance M. Smith, January 4, 1994, PECA.

Before he accepted the offer, Hendrix prescribed several conditions: he had to bolster his senior officer team; he wanted to impose his view of the organization and mission of Panhandle Eastern on the firm; and he had to be paid only in stock for a three year period. His income would be entirely dependent upon the performance of the company. The board accepted these terms and set his compensation at 100,000 shares of common per year. On November 12, Dennis R. Hendrix replaced Burguieres as president, CEO, and chairman.[34]

Hendrix reorganized Panhandle. He adopted a decentralized structure: "If a person is responsible for the operating unit and that unit's bottom line," he said, "they are going to be much more aggressive in terms of controlling costs of various support activities."[35] Panhandle now had two basic business units. The first, which he would oversee on an interim basis, consisted of Panhandle Eastern Pipe Line Co., Trunkline Gas Co., and Trunkline LNG. The second, to be managed by George Mazanec (who had worked for Hendrix in the 1980s), included Texas Eastern Transmission Company and Algonquin Gas Transmission.[36] Hendrix assigned "management teams to the four pipelines, and [gave] them the autonomy to figure out what needed to be done, make recommendations, and go try and get it done. . . . "[37]

Hendrix added two other former Texas Eastern associates to his team – Paul Anderson and Derrill Cody – and set out to improve Panhandle's performance as soon as possible.[38] The stock market had him on a short leash. He placed Anderson in charge of the important Panhandle Eastern–Trunkline busi-

---

[34] See Panhandle Eastern, *Minutes*, November 12, 1990; "Panhandle Eastern names Hendrix as Chief Officer," *The Wall Street Journal*, November 13, 1990; Anne Pearson, "Ill Chairman to Quit at Panhandle Eastern," *Houston Chronicle*, November 13, 1990; and James R. Norman, "Incentivized," *Forbes*, December 7, 1992.

[35] Dennis R. Hendrix, interview by Christopher J. Castaneda and Clarance M. Smith, January 4, 1994, PECA.

[36] Panhandle Eastern, news release, April 24, 1991.

[37] Dennis R. Hendrix, interview by Christopher J. Castaneda and Clarance M. Smith, January 4, 1994, PECA.

[38] Paul M. Anderson, interview by Christopher J. Castaneda and Clarance M. Smith, December 21, 1993, PECA.

Figure 10.1 Office of the CEO, 1993: Paul Anderson, Dennis Hendrix, and George Mazanec

ness unit and set out to improve the efficiency of an organization in which employees were not always sure what their specific duties were or what was expected of them. "There was not a focus," recalled Anderson, during 1989 and 1990, "almost the entire senior management of Panhandle disappeared." Leadership was needed if they were going to cut costs and improve performance quickly.[39]

Management took "a look at everything" and tried at once "to clean up everything that [could] be cleaned up."[40] The largest and most urgent problem was the LNG operation, and one of Hendrix's first activities was to conclude the revaluation of the Lake Charles facility. In December 1990, Panhandle Eastern took a $215 million, after tax write-down on the terminal. To capture the synergies of a Panhandle Eastern/Trunkline and Texas Eastern/Algonquin connection, the company had to employ its excess capacity in the northeast.[41] To do so, it increased its pipeline expansion expenditures by 10 percent in 1992 (to $275 million).[42]

One more turn on the regulatory wheel

Panhandle's reorganization was thus well under way when the FERC issued a new order – 636, on April 8, 1992 – that it hoped would at last solve the problems regulation had done so much to create in this industry. This order was the culmination of several others issued by FERC since the 1970s when the gas shortages heralded the end of the New Deal regulatory system. As FERC chairman Martin L. Allday explained: "Order No. 636 will signal the end of the seemingly endless transition period the gas industry has been in. . . . Order 636] is the next and hopefully last major step in the commission's efforts to allow competition rather than regulation to govern how pipelines function."[43] But in fact this order, like the others, stipu-

---

[39] Hendrix also brought in two new directors both of whom had previously been directors of Texas Eastern. These included Ralph O'Connor, chairman and CEO of a private investment firm and a former son-in-law of George R. Brown, one of Texas Eastern's founders and its chairman from 1947 through 1971, and Robert Cizik, chairman and president of Cooper Industries, Inc. Paul M. Anderson, interview by Christopher J. Castaneda and Clarance M. Smith, December 21, 1993, PECA.

[40] Dennis R. Hendrix, interview by Christopher J. Castaneda and Clarance M. Smith, January 4, 1994, PECA.

[41] Ibid.

[42] Robert Johnson, "Panhandle Eastern Prospers by Hauling Others' Fuel," *The Wall Street Journal*, February 2, 1993.

[43] See FERC, *Fact Sheet*, Remarks of FERC Chairman Allday, April 8, 1992.

Table 10.3. *Gas sales vs. transportation, 1983–1993 (Bcf)*

	Panhandle Eastern		Texas Eastern		Interstate natural gas delivered by carriage (%)
	Carriage(1)	Sales(2)	Carriage(3)	Sales(4)	
1983	–	677	158	1,164	5
1984	31	643	143	1,193	8
1985	167	515	204	792	17
1986	240	429	287	729	37
1987	567	286	437	662	56
1988	775	199	330	687	63
1989	900	216	673	657	71
1990	997	180	815	452	79
1991	1,016	148	959	250	84
1992	1,131	156	1,161	214	87
1993	1,117	88	1,281	35	90

Note: Panhandle Eastern and Texas Eastern only transported natural gas as of the fourth quarter of 1993.
Source: Moody's Public Utility Manual, 1987, 1990; INGAA, *"Carriage through 1993,"* Report no. 94–2, June, 1994; Texas Eastern, *1988 Financial and Statistical Summary*, p. 6; Panhandle Eastern, *Annual Report, 1993*, pp. 36–8.

lated what the structure of the industry and its major firms would be.

Order 636 mandated the "unbundling" of the traditional natural gas pipeline merchant service. This meant in effect that services including transportation, gathering, and storage could be controlled individually by contracts based on market forces rather than regulatory policy (Table 10.3). Order 636 also provided for the issuance of blanket "sales for resale" certificates for gas sales at market based prices and, most importantly, nondiscriminatory open-access transportation services, or contract carriage.[44] Pipelines would now compete by offering efficient and innovative transportation and marketing services to producers and consumers. Generally, pipelines made more money buying and selling gas than transporting it, but Order 636 prompted gas firms to significantly enlarge their marketing services, including purchasing, gathering, and processing gas. In 1994, Panhandle acquired Associated Natural Gas, Inc.

[44] Ibid.

Table 10.4. *PEC consolidated financial data, 1989–1994*

Year	Operating revenues	Net income	Earnings per share	Long-term debt	Total assets
1989	2,539	29	.41	2,786	6,266
1990	2,994	(286)	(3.21)	2,470	7,055
1991	2,459	86	.87	2,252	6,756
1992	2,434	187	1.73	2,479	6,946
1993	2,121	148	1.29	1,923	6,731
1994[a]	4,585	225	1.51	2,364	7,508

[a]1994 figures include data from Associated Natural Gas Corporation acquired in 1994.
Source: Panhandle Eastern Corporation, *Financial and Statistical Supplement to the 1991 Annual Report* and Panhandle Eastern, *Annual Report, 1983, 1993.*

in exchange for 28.4 million shares of common stock to augment its marketing capabilities (Table 10.4). The new federal pipeline policy marked the recognition by regulators that a large measure of competition was necessary for this industry to function successfully.[45]

In effect, however, Order 636 also ruled that this industry, unlike most others, would not benefit from vertical integration. In most of the nation's high-tech industries, firms had found economies by eliminating transaction costs and by linking the capacities for mass distribution and mass production. But fearful of the economic power this sort of integration would apparently bestow on the successful companies, the FERC compromised in a manner characteristic of political organizations. Market forces, Order 636 proclaimed, were absolutely essential – so long as they functioned within a structure specified carefully by the regulators who had in the past found it very difficult to keep their policies in tune with a rapidly changing economy. In brief, the new institutional set-

[45] In *Contrived Competition*, Richard Vietor uses El Paso Natural Gas as a case study in an historical analysis of natural gas regulation. Vietor claims that "When the court affirmed this rule [Order 500] in 1988, regulated competition became an established fact" (p. 161). The analysis of Panhandle Eastern's history through Order 636 suggests that regulated competition did not truly become a reality until this Order mandated open access, nondiscriminatory carriage.

ting was a halfway house, better for its partial acceptance of markets but on trial for its rigid stipulations about the nature of firms and their roles in the natural gas business.

The FERC's new order required all pipelines to restructure their services for the 1993–4 winter heating season. The first pipeline to do so was Transwestern (a former subsidiary of Texas Eastern) on February 1, 1993. Panhandle was, however, the first (May 1, 1993) of the large interstate lines to restructure under Order 636. It was this transition that prompted the meeting at Panhandle with which this book began. As you may recall, it was at that session that Paul Anderson wrote the words: "merchant service" on the blackboard and drew a diagonal line through the phrase. Texas Eastern and Algonquin offered restructured services as of June 1, and Trunkline Gas's restructuring became effective on September 1, 1993.[46]

Now Panhandle Eastern was a modified common carrier with a new customer base. The company's chief financial officer, Jay Hipple, explained: "We went from dealing with sales to 80 customers with $5 billion worth of revenues and now we've got $2.5 billion in revenues and we've probably got 5,000 customers. So the credit function takes on a bigger picture" (Table 10.5).[47] There were thousands of bills to collect each month and a new need for marketing capabilities. Panhandle moved quickly to develop these capabilities and improve its competitive position in the new environment. One of the new services was the Integrated Transportation Program (ITP), which used all of the company's interconnected pipelines to provide transportation of up to 261 MMcf/d for Northeastern customers. More flexible and customer-tailored services were developed using new technologies. Computer technology began to have a larger role in the new gas marketing scenario. Electronic bulletin boards allowed potential users of the transportation services to contract, nominate, schedule, allocate billing, and store gas.[48] LINK, a Customer Interface System, enabled transportation customers to

---

[46] "FERC Publishes Order No. 636 Compliance Effective Dates for 76 Pipelines," *Foster Natural Gas Report*, September 30, 1993, p. 4. Order 636 was subsequently modified by Orders 636-A and 636-B.

[47] James B. Hipple, interview by Christopher J. Castaneda and Clarance M. Smith, December 22, 1993, PECA.

[48] "The Brave New World of FERC Order 636 Opens New Frontiers," *PipeLines*, November–December 1993.

Table 10.5. *Composite financial information: U.S. gas
pipelines and distribution companies, 1981–1993 (millions)*

Year	Pipelines		Distributors	
	Total operating revenues	Net income	Total operating revenues	Net income
1981	$50,188	$3,047	$12,726	$361
1982	55,847	2,829	14,653	373
1983	53,577	2,749	16,743	521
1984	53,319	2,880	17,366	627
1985	45,738	1,757	21,510	831
1986	33,887	1,327	18,352	689
1987	27,275	1,015	16,513	713
1988	26,482	1,491	16,665	950
1989	23,883	2,270	18,761	1,097
1990	21,756	2,466	18,750	993
1991	19,818	955	17,812	961
1992	20,193	1,545	19,854	1,176
1993	20,061	2,865	21,810	2,017

Source: Moody's Public Utility Manual, 1994, p. a55; and 1995.

create or amend contracts and make on-line volume nominations using a personal computer.[49]

While guiding Panhandle Eastern successfully into this latest regulatory era, CEO Hendrix remains as concerned about the dynamics of the political process as are most U.S. business leaders. Regulatory systems, he says, reflect "the continual tension of one group of people to exert jurisdiction over another group of people. And it's a very delicate balance to make the thing work and make it produce the desired results." Knowing that "market forces are a necessary ingredient to an efficient industry," he and others are still worried about whether the United States has found the right "delicate balance."[50] Meanwhile they are riding what Paul Anderson sees as the industry's "last wave of expansion right now. We've got a billion dollars worth of projects on the drawing board that will carry us

[49] Panhandle Eastern, *Annual Report, 1992,* p. 9.
[50] Dennis R. Hendrix, interview by Christopher J. Castaneda and Clarance M. Smith, January 4, 1994, PECA.

Figure 10.2 Associated Natural Gas trading room

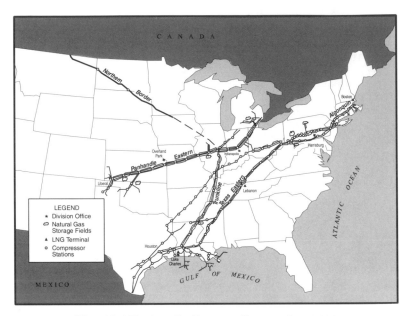

Map 10.1 Panhandle Eastern Corporation, 1993

through the next few years, but I think once those are in place that's it. There will be some fine tuning of the system . . . but I think the grid is there. The future is going to go to those who can use the grid more effectively and create services around the grid, very much like telecommunications."[51] Anderson and Hendrix have positioned Panhandle (Map 10.1), they believe, to be one of those which "can use the grid more effectively."

Reflections

The history of this one firm, from the 1920s to the 1990s, tells us much about the modern American business system and its links to a changing political economy. Panhandle was at first a relatively small, entrepreneurial enterprise in a fast-growing industry. In an era characterized by weak public controls of securities markets and energy firms, Panhandle found it difficult to crack the markets controlled by several large combines. Using a com-

[51] Paul M. Anderson, interview by Christopher J. Castaneda and Clarance M. Smith, December 21, 1993, PECA.

bination of economic and political power, the combines thwarted innovations and eventually swallowed this small business. Given that the nation needed the natural gas businesses like Panhandle were attempting to bring to midwestern and eastern markets, this outcome seemed to call for public interventions.

In the context of the Great Depression of the 1930s, intervention in several guises was forthcoming. William Maguire astutely exploited that new situation and antitrust policy to make Panhandle once again an independent and successful business. He did so despite the difficulties of engineering change in what was now a pervasively regulated industry. Maguire lost a number of his competitive efforts to expand Panhandle's operations in the midwest. He was, for instance, unable to provide direct service to large industrial customers and bypass the local utility. In the expansive economy of the forties and fifties, however, Panhandle was able to extend its operation in spite of these political impediments to entrepreneurship. This was, in fact, a golden age for the natural gas industry, which reaped great profits by linking new sources of cheap gas to fast-growing urban markets in the northeast, midwest, and far west.

Eventually, however, the dynamics of the regulatory process extended the reach of the New Deal controls, and the regulatory system began to choke the industry it was supposed to protect in the public interest. The regulatory institutions and their leaders characteristically thought in terms of maintaining short-term equilibria, satisfying stakeholders, and avoiding political crises. They were ill-suited to confronting the problems of long-term shifts in the energy economy.

But the economic transitions took place in spite of the government's efforts to resist them, and Panhandle Eastern was challenged to predict and accommodate to the changes that were taking place in both its markets and its regulatory setting. It had to do so as well without the firm hand of William Maguire at the tether. Like many other business organizations, Panhandle found the transition from entrepreneurial to managerial control difficult to make – even dangerous. The company's managers bet the firm on their ability to tap new sources of natural gas and to alleviate the nation's energy shortages in that manner. In particular, they made long-term contracts for large supplies of Algerian liquefied natural gas and invested heavily in the facilities needed to bring that gas to U.S. markets.

Figure 10.3 Panhandle Eastern Corporation board of directors, 1994 (seated left to right: Robert Cizik, George Mazanec, Dennis Hendrix, Paul Anderson, and Charles Duncan. Standing: Max Lents, Leo Linbeck, Lloyd Bentsen, Ann Maynard Gray, Harold Hook, Milton Carroll, Harry Ekblom, Ralph O'Connor, Cortlandt Dietler, and William Esrey)

Deregulation and the new supplies of gas that it brought forth undercut Panhandle's strategy and left the company vulnerable to a takeover. In this case, however, the market for corporate control worked indirectly to make Panhandle a more efficient and successful company over the long-term. The firm's managers fought off one takeover attempt and managed to acquire Texas Eastern, a pipeline that opened growing northeastern markets to the Panhandle system. Slimmed down after spinning off its diversified operations, Panhandle Eastern was able, after one more challenging bout with the market, to restructure and to become a profitable, growing business. It be-

Table 10.6. *PEC common stock price range and dividends paid, 1988–1994*

Annual quarter	Common stock		Dividends Paid
	High	Low	
Q188	27 3/8	21	$0.50
Q288	26 1/4	22 5/8	.50
Q388	27	23 1/2	.50
Q488	27 1/8	24 1/4	.50
Q189	27	20 1/2	.50
Q289	23 7/8	20 1/2	.50
Q389	29 3/8	23 1/2	.50
Q489	30 3/4	25 3/4	.50
Q190	29 3/4	26 1/2	.50
Q290	28 1/2	24 3/8	.50
Q390	24 1/2	13	.20
Q490	15	10 3/8	.20
Q191	15 1/2	9 7/8	.20
Q291	14 3/4	10 3/4	.20
Q391	13 7/8	10 1/4	.20
Q491	16 1/2	12 1/2	.20
Q192	15 7/8	13 1/2	.20
Q292	17 1/4	12 7/8	.20
Q392	19 3/8	15 3/4	.20
Q492	19 3/4	16 1/4	.20
Q193	23 3/4	16 3/4	.20
Q293	25	19 3/4	.20
Q393	27 1/4	23	.20
Q493	26 3/4	20 3/8	.20
Q194	25 1/2	20 5/8	.20
Q294	22 1/4	18 1/4	.20
Q394	23 1/2	19 1/2	.21
Q494	23 5/8	19 1/2	.21

Source: Panhandle Eastern, *Annual Report*, 1988–1994.

came a business that seems likely to be successful adapting to the new U.S. political economy, with its unique blend of market forces and the remnants of a New Deal regulatory system. The stock market was by this time confirming that prognosis: as Table 10.6 shows, the value of Panhandle Eastern's had more than doubled since the low point of the early 1990s.

APPENDIX 1

An earlier effort

In 1955, the Indiana Historical Society administered a grant from Panhandle Eastern to finance the research and writing of a history of the firm. Lana Ruegamer, the author of the Historical Society's institutional autobiography, noted that, "The project [Panhandle Eastern history] was to be a model for a series on Indiana business history, but unfortunately, like many pilot studies, it failed. Despite the efforts of several researchers, editors, and a supervisory board, the difficulties of the task proved insurmountable."[1] Robert B. Eckles, then an associate professor of History at Purdue University, worked on the project from 1955 through 1963.[2] In the early phase of his research, Eckles discussed the project with eminent business historians. He travelled to the Harvard Business School and conferred with Arthur H. Cole, business historian and librarian of the

---

[1] Lana Ruegamer, *A History of the Indiana Historical Society: 1830–1980* (Indianapolis: Indiana Historical Society, 1980), 200.

[2] Two of Eckles's published books are Robert B. Eckles, *Purdue Pharmacy: The First Century* (West Lafayette, IN: Purdue University, 1974); and Robert B. Eckles, *The Dean: A Biography of A. A. Potter* (West Lafayette, IN: Purdue University, 1974). Also see Robert B. Eckles, "What's Ahead for Natural Gas Regulation," *Public Utilities Fortnightly*, October 27, 1966, pp. 41–5.

Baker Memorial Library.[3] Eckles also met with Henrietta Larson who was in the process of writing *History of Humble Oil & Refining* and Ralph Hidy, author of a history of Standard Oil of New Jersey. Hidy recommended that Eckles focus on Panhandle Eastern's "financial structure" and "the relationships of the transmission line to local authorities, county, and state. . . ."[4]

Eckles involved graduate students in the project to research and write a history of Panhandle Eastern, and his research became a significant part of his seminars. Two doctoral students who assisted in the research for the Panhandle manuscript subsequently wrote doctoral dissertations on different aspects of the natural gas industry: Ralph W. Helfrich, Jr., "Administrative Regulation of Natural Gas Rates, 1898–1938" (University of Indiana, diss., 1961), and Earl D. Bragdon, "The Federal Power Commission and the Regulation of Natural Gas: a Study in Administrative and Judicial History" (University of Indiana, diss., 1962). These students were ultimately more successful than Eckles in publishing, albeit in dissertation form, their natural gas research.

Eckles decided to write a generally internal history of Panhandle Eastern with a concentration on finance. He was forced to depend a great deal on William G. Maguire, Panhandle's chairman, for information and access to records. Eckles wrote his concerns about his relationships with Maguire to the Indiana Historical Society:

> Mr. Maguire has assured me that he will be reticent about nothing – but I have discovered in the pipeline business, sabotage, espionage, and plain and fancy skullduggery has been the order of the day. Therefore, Mr. Maguire and other members of his executive staff may be embarrassingly silent about certain of their affairs. . . . The project history will lack a very great deal and can be weak if we cannot delve into the reasons for the

[3] See Arthur Cole, *Business Enterprise in its Social Setting* (Cambridge, Mass.: Harvard University Press, 1959) and Arthur Cole, *The Birth of a New Social Science Discipline: Achievements of the First Generation of American Economic and Business Historians, 1893–1974* (New York: Economic History Association, 1974).

[4] Robert B. Eckles, "Report of March, 1956," pp. 2, 8. Also see Henrietta Larson and Kenneth Wiggins Porter, *History of Humble Oil & Refining: A Study in Industrial Growth* (New York: Harper and Row, 1959); and Ralph Hidy, *Pioneering Big Business: The Standard Oil Company* (Boston, 1957).

important decisions and try to find out why they were made, as they were made, and when they were made. This is one of the hurdles or stumbling blocks that we must overcome.[5]

In another letter to his editors at the Indiana Historical Society, Eckles described Maguire as a "vindictive, capricious, dictatorial genius."[6] As one example of this behavior which affected his own work, Eckles noted that when he asked company vice-president Frederick H. Robinson, second-in-command and Maguire's son-in-law, for a company organization chart, Robinson "laughed for five minutes and kicked me out of his office."[7] This episode underscored Maguire's reputation for running the company with an iron fist, leaving few important decisions for his subordinates. Thus, Eckles noted that "Maguire cannot stand to be refuted by his top executives . . . That in personal relations he has been extraordinarily kind is also a fact. But those who are closest to him get kicked in the teeth regularly."[8] Also, Maguire was not pleased that Eckles expressed some admiration for the firm's deposed founder, Frank Parish.

Eckles, Helfrich (who became a member of the IHS board), and Bragdon took turns working on the manuscript through the early 1960s. Their efforts resulted in a fairly disorganized and disconnected set of heavily detailed chapters. Then, in an effort to salvage the manuscript, Luke Scheer, who served as one of Mo–Kan's public relations officers in the 1920s and 1930s, dramatically revised the manuscript. Scheer, who also planned on writing a history of the natural gas industry, worked on the manuscript through at least 1975 with no better results. Neither the Eckles nor the Scheer manuscripts were ever published.

Several versions of the original Eckles manuscript are located at the Indiana Historical Society in Indianapolis. Eckles's work proved useful to us since he had access to both people and sources no longer extant, and his drafts contained some important information we had not previously discovered.

---

[5] Robert B. Eckles, "Report, March 1956," p. 8, PECA.

[6] Robert B. Eckles to Miss Gayle Thornborough and J. A. Batchelor, July 28, 1959, IHS.

[7] Ibid.

[8] Ibid.

APPENDIX 2

Interview list

Ables, Dorothy	01/31/92
Alholm, John	02/10/92
Allen, Milt	02/11/93
Allison, Robert J.	12/20/93
Anderson, Page	04/14/92
Anderson, Paul M.	12/21/93
Auvenshine, W. L.	04/29/92
Barnard, Marla	09/02/93
Becker, Joe	08/14/91
Bond, Ed	08/29/91
Boyter, Lee	04/20/92
Broussard, George	08/22/91
Campbell, J. Del	09/10/91
Cheatham, Jack	01/13/94
Dillon, Bob	08/25/92
Dufield, Tom	10/01/93
Duff, E. L.	11/04/91
Dunsworth, Dunny	11/22/91
Ebdon, Fred	09/16/91
Fordemwalt, Andy	08/25/92
Fowler, Fred	02/12/92
Hamilton, Kenny	08/19/91
Hanna, Larry	08/24/92
Harmond, Ron	09/08/93

Hart, James W., Jr.	12/17/93
Henderson, Deep	01/24/92
Hendrix, Dennis R.	01/04/94
Hennig, Dan	11/22/91
Hermes, Malcolm	09/17/91
Heying, Dale	04/11/92
Hinton, Charlie	06/30/93
Hipple, James B.	12/22/93
Hughes, Dave	11/21/91
Hunsucker, Robert D.	10/01/93
Irwin, John A.	09/06/91
Irwin, Thomas B.	03/18/93
Johnson, Jerry	06/08/92
Kalen, Kenny	08/25/92
Kennedy, Truitt	04/07/92
Kurk, George	04/14/92
Laforce, John	12/02/93
Levin, Arnold J.	01/08/93
McGrath, Jerry	01/13/94
Manley, Charles G.	12/20/93
Mazanec, George	01/14/94
Michael, Kirk	09/28/93
Nelson, Ernie	04/24/92
O'Shields, Richard L.	04/08/92
Parent, Len	09/05/91
Parks, Howell	04/30/92
Perry, Wayne	07/01/93
Price, J. M. "Mack"	09/10/93
Reed, Bob	10/14/91
Rehak, Bob	08/11/93
Rigdon, Vern	08/26/92
Rist, Barry	08/25/92
Roberts, Joe	09/09/91
Robertson, Larry	09/17/91
Sanders, Bertha	10/10/91
Schomaker, John F.	04/22/92
Schulze, H. E.	08/09/91
Seely, Dwight	09/19/91
Segura, Roger	10/07/93
Shibley, Raymond M.	01/13/94
Sieger, John A.	12/21/93
Smith, Cy	09/12/91

Taylor, Paul A.	12/20/93
Townsend, John	12/30/93
Turbeville, Tony	09/27/93
Van Leuvan, Robert A.	03/16/93
Welch, Harry S.	04/09/92
Wellspring, Tom	10/17/91
Zangara, Evelyn	08/11/91

Group Interview: Secretaries, 10/04/91
Hardy, Leona
Kibler, Phyllis
Lokey, Elizabeth
Merritt, Pat
Smith, Morna
Wright, Valera

Group Interview: Retirees in Liberal, Kansas, 07/01/93
Handley, Ray
Hanna, John
Hathaway, Hoppy
Jackson, Johnny
Leaming, Ray
Massey, Harry
Nethercote, Sam
Rowley, Hank
Thatcher, Earl
Thomas, Paul

Index